CLIMATE CHANGE AND ITS CAUSES, EFFECTS AND PREDICTION SERIES

EFFECTS OF CLIMATE CHANGE ON ENERGY PRODUCTION AND USE IN THE U. S.

CLIMATE CHANGE AND ITS CAUSES, EFFECTS AND PREDICTION SERIES

Global Climate Change
Horace M. Karling (Editor)
2001. ISBN: 1-56072-999-6

Global Climate Change Revisited
Harace B. Karling (Editor)
2007. ISBN: 1-59454-039-X

Climate Change Research Progress
Lawrence N. Peretz (Editor)
2008. ISBN: 1-60021-998-5

Climate Change: Financial Risks
United States Government Accountability Office
2008. ISBN: 978-1-60456-488-4

Post-Kyoto: Designing the Next International Climate Change Protocol
Matthew Clarke
2008. ISBN: 978-1-60456-840-0

Post-Kyoto: Designing the Next International Climate Change Protocol
Matthew Clarke
2008. ISBN: 978-1-61668-101-2
(Online Book)

Economics of Policy Options to Address Climate Change
Gregory N. Bartos
2009. ISBN: 978-1-60692-116-6

The Effects of Climate Change on Agriculture, Land Resources, Biodiversity in the United States
Peter Backlund, Anthony Janeto and David Schimel
2009. ISBN: 978-1-60456-989-6

Thresholds of Climate Change in Ecosystems
Vicente Orostegui (Editors)
2009. ISBN: 978-1-60741-487-2

Climate Variability, Modeling Tools and Agricultural Decision-Making
Angel Utset (Editor)
2009. ISBN: 978-1-60692-703-8

Climate Variability, Modeling Tools and Agricultural Decision-Making
Angel Utset (Editor)
2009. ISBN: 978-1-60876-791-5
(Online Book)

Carbon Offsets: Examining their Role in Greenhouse Gas Reduction
Karen T. Morningstar (Editor)
2010. ISBN: 978-1-60741-444-5

Constructing Climate Change Legislation: Background and Issues
Gerald P. Overhauser (Editor)
2009. ISBN: 978-1-60692-986-5

Emissions Trading: Lessons Learned from the European Union and Kyoto Protocol Climate Change Programs
Ervin Nagy and Gisella Varga (Editors)
2009. ISBN: 978-1-60741-194-9

Designing Greenhouse Gas Reduction and Regulatory Systems
Sonja Enden (Editor)
2009. ISBM: 978-1-60741-195-6

Global Climate Change: International Perspectives and Responses
Elias D'Angelo (Editor)
2009. ISBN: 978-1-60741-233-5

Disputing Global Warming
Anton Horvath and Boris Molnar (Editors)
2009. ISBN: 978-1-60741-235-9

Disputing Global Warming
Anton Horvath and Boris Molnar (Editors)
2009. ISBN: 978-1-60876-503-4
(Online Book)

Focus on Climate Change and Health
Viroj Wiwanitkit
2009. ISBN: 978-1-60741-247-2

Land Use and Climate Change
Suresh C. Rai
2009. ISBN: 978-1-60741-362-2

Impacts of Climate Change on Transportation and Infrastructure A Gulf Coast Study
Iason Pavlopoulos (Editors)
2009. ISBN: 978-1-60741-424-7

Coastal Sensitivity to Sea Level Rise Focusing on the Mid-Atlantic Region
Melvin C. Urajner (Editors)
2009. ISBN: 978-1-60741-440-7

Lightning in the Tropics: From a Source of Fire to a Monitoring System of Climatic Changes
Osmar Pinto, Jr.
2009. ISBN: 978-1-60741-764-4

Carbon Capture and Greenhouse Gases
Imrus Juhász and Gyorgy Halász (Editors)
2010. ISBN: 978-1-60692-089-3

The Science of Climate Change and Policy Implications
Kyle S. Hartzell (Editor)
2010. ISBN: 978-1-60741-448-3

Effects of Climate Change on Energy Production and Use in the U.S.
Alrik M. Solberg (Editor)
2010. ISBN: 978-1-60741-426-1

Policy Option Issues for CO2 Emissions
Nikolaus Vogler (Editors)
2010. ISBN: 978-1-60741-381-3

The Role of Agriculture in Carbon Capture and Climate Change
Arvid Bjurstrom (Editors)
2010. ISBN: 978-1-60741-445-2

Carbon Sequestration: Methods, Modeling and Impacts
Elke Hoch and Siegbert Grunwald (Editors)
2010. ISBN: 978-1-60741-498-8

Climate Change Litigation and Law
Jean-François Masson (Editor)
2010. ISBN: 978-1-60876-089-3

Carbon Tax and Cap-and-Trade Tools: Market-Based Approaches for Controlling Greenhouse Gases
Nelson E. Burney (Editor)
2010. ISBN: 978-1-60876-137-1

Global Change and Forestry: Economic and Policy Impacts and Responses
Jianbang Gan Stephen Grado and Ian. A. Munn (Editor)
2010. ISBN: 978-1-60876-262-0

Climate Change Measures and Trade Considerations
George R. Fried (Editor)
2010. ISBN: 978-1-60876-756-4

Assessing Climate Change Impacts on the United States
Alex B. McNeill (Editor)
2010. ISBN: 978-1-60876-160-9

Assessing Climate Change Impacts on the United States
Alex B. McNeill (Editor)
2010. ISBN: 978-1-60876-160-9
(Online Book)

CLIMATE CHANGE AND ITS CAUSES, EFFECTS AND PREDICTION SERIES

EFFECTS OF CLIMATE CHANGE ON ENERGY PRODUCTION AND USE IN THE U. S.

ALRIK M. SOLBERG
EDITOR

Nova Science Publishers, Inc.
New York

For permission to use material from this book please contact us:
Telephone 631-231-7269; Fax 631-231-8175
Web Site: http://www.novapublishers.com

LIBRARY OF CONGRESS CATALOGING-IN-PUBLICATION DATA

Effects of climate change on energy production and use in the U.S. / editor, Alrik M. Solberg.
p. cm.
Includes bibliographical references and index.
ISBN 978-1-60741-426-1 (softcover)
1. Energy consumption--Environmental aspects--United States. 2. Energy consumption--Climatic factors--United States. 3. Energy industries--Environmental aspects--United States. 4. Environmental policy--United States. I. Solberg, Alrik M.
HD9502.U52E386 2009
333.790973--dc22
2009035596

Published by Nova Science Publishers, Inc. † New York

Contents

PREFACE

This book summarizes what is currently known about effects of climate change on energy production and use in the United States. It focuses on three questions: 1)How might climate change affect energy consumption in the U.S. 2) How might climate change affect energy production and supply in the U.S. and 3) How might climate change have other effects that indirectly shape energy production and consumption in the U.S. Generally, it is important to be careful about answering these questions for two reasons. One reason is that the available research literature on many of the key issues are limited, supporting a discussion of issues but not definite conclusions about answers. A second reason is that, as with many other categories of climate change effects in the U.S., the effects depend on more than climate change alone, such as patterns of economic growth and land use, patterns of population growth and distribution, technological change, and social and cultural trends that could shape policies and actions, individually and institutionally. This book concludes that, based on what we know now, there are reasons to pay close attention to possible climate change impacts on energy production and use and to consider ways to adapt to possible adverse impacts and take advantage of possible positive impacts.

In: Effects of Climate Change on Energy Production... ISBN: 978-1-60741-426-1
Editor: Alrik M. Solberg

Chapter 1

EXECUTIVE SUMMARY*

U.S. Climate Change Science Program

Climate change is expected to have noticeable effects in the United States: a rise in average temperatures in most regions, changes in precipitation amounts and seasonal patterns in many regions, changes in the intensity and pattern of extreme weather events, and sea level rise. Some of these effects have clear implications for energy production and use. For instance, average warming can be expected to increase energy requirements for cooling and reduce energy requirements for warming. Changes in precipitation could affect prospects for hydropower, pos itively or negatively. Increases in storm intensity could threaten further disruptions of the sorts experienced in 2005 with Hurricane Katrina. Concerns about climate change impacts could change perceptions and valuations of energy technology alternatives. Any or all of these types of effects could have very real meaning for energy policies, decisions, and institutions in the United States, affecting discussions of courses of action and appropriate strategies for risk management.

This report summarizes what is currently known about effects of climate change on energy production and use in the United States. It focuses on three questions, which are listed below along with general short answers to each. Generally, it is important to be careful about answering these questions for two

* This is an edited, reformatted and augmented version of a U. S. Climate Change Science Program publication dated February 2008.

reasons. One reason is that the available research literatures on many of the key issues are limited, supporting a discussion of issues but not definite conclusions about answers. A second reason is that, as with many other categories of climate change effects in the U.S., the effects depend on more than climate change alone, such as patterns of economic growth and land use, patterns of population growth and distribution, technological change, and social and cultural trends that could shape policies and actions, individually and institutionally.

The report concludes that, based on what we know now, there are reasons to pay close attention to possible climate change impacts on energy production and use and to consider ways to adapt to possible adverse impacts and take advantage of possible positive impacts. Although the report includes considerably more detail, here are the three questions along with a brief summary of the answers:

- **How might climate change affect energy consumption in the United States?** The research evidence is relatively clear that climate warming will mean reductions in total U.S. heating requirements and increases in total cooling requirements for buildings. These changes will vary by region and by season, but they will affect household and business energy costs and their demands on energy supply institutions. In general, the changes imply increased demands for electricity, which supplies virtually all cooling energy services but only some heating services. Other effects on energy consumption are less clear.

- **How might climate change affect energy production and supply in the United States?** The research evidence about effects is not as strong as for energy consumption, but climate change could affect energy production and supply (a) if extreme weather events become more intense, (b) where regions dependent on water supplies for hydropower and/or thermal power plant cooling face reductions in water supplies, (c) where temperature increases decrease overall thermoelectric power generation efficiencies, and (d) where changed conditions affect facility siting decisions. Most effects are likely to be modest except for possible regional effects of extreme weather events and water shortages.

- **How might climate change have other effects that indirectly shape energy production and consumption in the United States?** The research evidence about indirect effects ranges from abundant information about possible effects of climate change policies on energy technology choices to extremely limited information about such issues as effects on energy security. Based on this mixed evidence, it appears that climate change is likely to affect risk management in the investment behavior of some energy institutions, and it is very likely to have some effects on energy technology R&D investments and energy resource and technology choices. In addition, climate change can be expected to affect other countries in ways that in turn affect U.S. energy conditions through their participation in global and hemispheric energy markets, and climate change concerns could interact with some driving forces behind policies focused on U.S. energy security.

Because of the lack of research to date, prospects for adaptation to climate change effects by energy providers, energy users, and society at large are speculative, although the potentials are considerable. It is possible that the greatest challenges would be in connection with possible increases in the intensity of extreme weather events and possible significant changes in regional water supply regimes. But adaptation prospects depend considerably on the availability of information about possible climate change effects to inform decisions about adaptive management, along with technological change in the longer term.

Given that the current knowledge base is so limited, this suggests that expanding the knowledge base is important to energy users and providers in the United States. Examples of research priorities – which call for contributions by a wide range of partners in federal and state governments, industry, nongovernmental institutions, and academia – are identified in the report.

In: Effects of Climate Change on Energy Production... ISBN: 978-1-60741-426-1
Editor: Alrik M. Solberg

Chapter 2

INTRODUCTION

U.S. Climate Change Science Program

As a major expression of its objective to provide the best possible scientific information to support decision-making and public discussion on key climaterelated issues, the U.S. Climate Change Science Program (CCSP) has commissioned 21 "synthesis and assessment products" (SAPs) to summarize current knowledge and identify priorities for research, observation, and decision support in order to strengthen contributions by climate change science to climate change related decisions.

These reports arise from the five goals of CCSP (*http://www.*Climate science. gov), the fourth of which is to "understand the sensitivity and adaptability of different natural and managed ecosystems and human systems to climate and related global changes." One of the seven SAPs related to this particular goal is concerned with analyses of the effects of global change on energy production and use (SAP 4.5). The resulting SAP, this report, has been titled "Effects of Climate Change on Energy Production and Use in the United States."

This topic is relevant to policy-makers and other decision-makers because most discussions to date of relationships between the energy sector and responses to concerns about climate have been very largely concerned with roles of energy production and use in climate change mitigation. Along with these roles of the energy sector as a *driver* of climate change, the energy sector is also subject to *effects* of climate change; and these possible effects – along with adaptation strategies to reduce any potential negative costs from them – have received much less attention. For instance, the U.S. National Assessment of Possible Consequences of Climate Variability and Change (NACC, 2001) considered effects on five sectors, such as water and health; but energy was not one of those sectors, even though the Global Change Research Act of 1990 had listed energy as one of several sectors of particular interest.

Because the topic has not been a high priority for research support and institutional analysis, the formal knowledge base is in many ways limited. As a starting point for discussion, this product compiles and reports what is known about likely or possible effects of climate change on energy production and use in the United States, within a more comprehensive framework for thought about this topic, and it identifies priorities for expanding the knowledge base to meet needs of key decision-makers.

1.1. Background

Climate change is expected to have certain effects in the United States: a rise in average temperatures in most regions, changes in precipitation amounts and seasonal patterns in many regions, changes in the intensity and pattern of extreme weather events, and sea level rise [(IPCC, 2001a; NACC, 2001; also see other SAPs, including 2.1b and 3.2)].

Some of these effects have clear implications for energy production and use. For instance, average warming can be expected to increase energy requirements for cooling and reduce energy requirements for warming. Changes in precipitation could affect prospects for hydropower, positively or negatively. Increases in storm intensity could threaten further disruptions of the sorts experienced in 2005 with Hurricane Katrina. Concerns about climate change impacts could change perceptions and valuations of energy technology alternatives. Any or all of these types of effects could have very real meaning for energy policies, decisions, and institutions in the United States, affecting discussions of courses of action and appropriate strategies for risk management.

According to CCSP, an SAP has three end uses: (1) informing the evolution of the research agenda; (2) supporting adaptive management and planning; and (3) supporting policy formulation. This product will inform policymakers, stakeholders, and the general public about issues associated with climate change implications for energy production and use in the United States, increase awareness of what is known and not yet known, and support discussions of technology and policy options where the knowledge base is still at an early stage of development.

The central questions addressed by SAP 4.5 follow:

- How might climate change affect energy consumption in the United States?
- How might climate change affect energy production and supply in the United States?
- How might climate change affect various contexts that indirectly shape energy production and consumption in the United States, such as energy technologies, energy institutions, regional economic growth, energy prices, energy security, and environmental emissions?

SAP 4.5 is being completed by the end of the second quarter of calendar year 2007 (June 30, 2007), following a number of steps required for all SAPs in scoping the study, conducting it, and reviewing it at several stages (see the section below on How the Report Was Developed).

1.2. The Topic of This Synthesis and Assessment Report

This report summarizes the current knowledge base about possible effects of climate change on energy production and use in the United States as a contributor to further studies of the broader topic of effects of global change on energy production and use. It also identifies where research could reduce uncertainties about vulnerabilities, possible effects, and possible strategies to reduce negative effects and increase adaptive capacity and considers priorities for strengthening the knowledge base. As is the case for most of the SAPs, it does not include new analyses of data, new scenarios of climate change or impacts, or other new contributions to the knowledge base, although its presentation of a framework for thought about energy sector impacts is in many ways new.

As indicated above, the content of SAP 4.5 includes attention to the following issues:

- possible effects (both positive and negative) of climate change on energy *consumption* in the United States (Chapter 2),

- possible effects (both positive and negative) on energy *production and supply* in the United States (Chapter 3), and
- possible *indirect effects* on energy consumption and production (Chapter 4).

These chapters are followed by a final chapter that provides conclusions about what is currently known, prospects for adaptation, and priorities for improving the knowledge base.

1.3. Previous Assessments of This Topic

As mentioned on page 1, unlike some of the other sectoral assessment areas identified in the Global Change Research Act of 1990—such as agriculture, water, and human health—energy was not the subject of a sectoral assessment in the *National Assessment of Possible Consequences of Climate Variability and Change*, completed in 2001 (NACC, 2001). As a result, SAP 4.5 draws upon a less organized knowledge base than these other sectoral impact areas. On the other hand, by addressing an assessment area not covered in the initial national assessment, SAP 4.5 will provide new information and perspectives.

The subject matter associated with SAP 4.5 is incorporated in two chapters of the Working Group II contribution to the Intergovernmental Panel on Climate Change (IPCC) Fourth Assessment Report (Impacts, Adaptation, and Vulnerability), scheduled for completion in 2007. Chapter 7, "Industry, Settlement, and Society," Section 7.4.2.1, briefly summarizes the global knowledge base about possible impacts of climate change on energy production and use, reporting relevant research from the United States but not assessing impacts on the United States. Chapter 14, "North America," summarizes the knowledge base about possible impacts of climate change in this continent, including the United States, in Sections 14.2.8 and 14.4.8.

1.4. How the Report Was Developed

SAPs are developed according to guidelines established by CCSP based on processes that are open and public. These processes include a number of steps before approval to proceed, emphasizing both stakeholder participation and CCSP reviews of a formal prospectus for the report, a number of review steps including both expert reviewers and public comments, and final reviews by the CCSP Interagency Committee and the National Science and Technology Council (NSTC).

The process for producing the report was focused on a survey and assessment of the available literature, in many cases including documents that were not peer-reviewed but the authors determined to be valid, using established analytic-deliberative practices. It included identification and consideration of relevant studies carried out in connection with CCSP, the Climate Change Technology Program (CCTP), and other programs of CCSP agencies (e.g., the Energy Information Administration), and consultation with stakeholders such as the electric utility and energy industries, environmental non-governmental organizations, and the academic research community to determine what analyses have been conducted and reports have been issued. Where quantitative research results are limited, the process considers the degree to which qualitative statements of possible effects may be valid as outcomes of expert deliberation, utilizing the extensive review processes built into the SAP process to contribute to judgments about the validity of the statements.

Box 1.1 SAP 4.5 Author Team

Thomas J. Wilbanks	*Oak Ridge National Laboratory, Coordinator*
Vatsal Bhatt	*Brookhaven National Laboratory*
Daniel E. Bilello	*National Renewable Energy Laboratory*
Stanley R. Bull	*National Renewable Energy Laboratory*
James Ekmann	*National Energy Technology Laboratory*
William C. Horak	*Brookhaven National Laboratory*
Y. Joe Huang	*Lawrence Berkeley National Laboratory*
Mark D. Levine	*Lawrence Berkeley National Laboratory*
Michael J. Sale	*Oak Ridge National Laboratory*
David K. Schmalzer	*Argonne National Laboratory*
Michael J. Scott	*Pacific Northwest National Laboratory*
Sherry B. Wright	*Oak Ridge National Laboratory, Administrative Coordinator*

SAP 4.5 is authored by staff from the Department of Energy (DOE) national laboratories, drawing on their own expertise and knowledge bases and also upon other knowledge bases, including those within energy corporations and utilities, consulting firms, nongovernmental organizations, state and local governments,

and the academic research community. DOE has assured that authorship by DOE national laboratory staff will in no way exclude any relevant research or knowledge, and every effort is being made to identify and utilize all relevant expertise, materials, and other sources. For the author team of SAP 4.5, see Box 1.1.

Stakeholders participated during the scoping process, have provided comments on the prospectus, and submitted comments on the product during a public comment period, as well as other comments via the SAP 4.5 web site. The development of SAP 4.5 included active networking by authors with centers of expertise and stakeholders to assure that the process was fully informed about their knowledge bases and viewpoints.

1.5. How to Use This Report

The audience for SAP 4.5 includes scientists in related fields, decision-makers in the public sector (federal, state, and local governments), the private sector (energy companies, electric utilities, energy equipment providers and vendors, and energy-dependent sectors of the economy), energy and environmental policy interest groups, and the general public. Even though this report is unable—based on existing knowledge—to answer all relevant questions that might be asked by these interested parties, the intent is to provide information and perspectives to inform discussions about the issues and to clarify priorities for research to reduce uncertainties in answering key questions.

As indicated above, because of limitations in available research literatures, in some cases the report is only able to characterize categories of possible effects without evaluating what the effects are likely to be. In other cases, the report offers preliminary judgments about effects, related to degrees of likelihood: likely (2 chances out of 3), very likely (9 chances out of 10), or virtually certain (99 chances out of 100).

This report avoids the use of highly technical terminology, but a glossary and list of acronyms are included at the end of the report.

In: Effects of Climate Change on Energy Production… ISBN: 978-1-60741-426-1
Editor: Alrik M. Solberg

Chapter 3

EFFECTS OF CLIMATE CHANGE ON ENERGY USE IN THE UNITED STATES

U.S. Climate Change Science Program

2.1. INTRODUCTION

As the climate of the world warms, the consumption of energy in climate-sen¬sitive sectors is likely to change. Possible effects include (1) decreases in the amount of energy consumed in residential, commercial, and industrial buildings for space heating and increases for space cooling; (2) decreases in energy used directly in certain processes such as residential, commercial, and industrial water heating, and increases in energy used for residential and commercial refrigera¬tion and industrial process cooling (e.g., in thermal power plants or steel mills); (3) increases in energy used to supply other resources for climate-sensitive processes, such as pumping water for irrigated agriculture and municipal uses; (4) changes in the balance of energy use among delivery forms and fuel types, as between electricity used for air conditioning and natural gas used for heating; and (5) changes in energy consumption in key climate-sensitive sectors of the economy, such as transportation, construction, agriculture, and others.

In the United States, some of these effects of climate change on energy consumption have been studied enough to produce a body of literature with empirical results. This is especially the case for energy demand in residential and commercial buildings, where studies of the effects of climate change have been under way for about 20 years. There is very little literature on the other effects mentioned above.

This chapter summarizes current knowledge about potential effects of climate change on energy demand in the United States. The chapter mainly focuses on the effects of climate change on energy consumption in buildings (emphasizing space heating and space cooling, but also addressing net energy use, peak loads, and adaptation), because studies of these effects account for most of the available knowledge. The chapter more briefly addresses impacts of climate change on energy use in other sectors, including transportation, construction, and agriculture, where studies are far less available. The final section summarizes the chapter's conclusions.

2.2. ENERGY CONSUMPTION IN BUILDINGS

2.2.1. Overview

U.S. residential and commercial buildings currently use about 20 quadrillion Btus (quads) of delivered energy per year (equivalent to about 38 quads of primary energy, allowing for electricity production-related losses). This energy consumption accounts directly or indirectly for 0.6 GT of carbon emitted to the atmosphere (38% of U.S. total emissions of 1.6 GT and approximately 9% of the world fossil-fuel related anthropogenic emissions of 6.7 GT (EIA, 2006). The U.S. Energy Information Administration (EIA) has projected that residential and commercial consumption of delivered energy would increase to 26 quads (53 quads primary energy) and corresponding carbon emissions to 0.9 GT by the year 2030 (EIA, 2006). However, these routine EIA projections do not account for the effects any temperature increases on building energy use that may occur as a result of global warming, nor do they account for consumer reactions to a warmer climate, such as an increase in the adoption of air conditioning.

To perform an assessment of the impact of climate change on energy demand, it is helpful to have as context a set of climate scenarios. The Intergovernmental Panel on Climate Change (IPCC) projected in 2001 that climate could warm relative to 1990 by 0.4°C to 1.2°C by the year 2030 and by 1.4°C to 5.8°C by the end of the 21st century (Cubasch et al., 2001 and Ruosteenoja, et al. 2003) performed a reanalysis of the seventeen 2001 IPCC climate simulations by seven different climate models at the regional level. Their results for the United States are reported for three subregions, four seasons, and three major time steps, as summarized in Table 2.1. This is not the only set of climate scenarios available, and the energy studies cited in this chapter often use other scenarios; but the table broadly characterizes the range of average temperature changes that might occur in the United States in the 21st century and can provide context for the various energy impact analyses that have been done.

Approximately 20 studies have been done since about 1990 concerning the effect of projected climate change on energy consumption in residential and commercial buildings in the United States. Some of these studies concern particular states or regions, and the impacts estimated depend crucially on local conditions.

Table 2.1. Seasonal Temperature Increases for Three U.S. Regions (°C) In Winter (DJ F), Spring (MAM), Summer (JJA), and Fall (SON). Derived From Ruosteenoja et al., 2003.

Region and Season	TIME STEP					
	2010-2039 (2020)		2040-2069 (2050)		2070-2099 (2080)	
	Median	Range	Median	Range	Median	Range
WESTERN U.S.						
DJF	1.6	0.5-2.4	2.3	1.0-4.2	4.1	2.0-7.6
MAM	1.4	0.5-1.9	2.5	1.1-4.1	3.8	1.0-7.6
JJA	1.8	0.8-2.6	2.8	1.7-5.2	4.2	2.8-9.1
SON	1.3	0.5-2.1	2.8	1.4-4.6	3.9	1.6-8.0
CENTRAL U.S.						
DJF	1.6	0.0-2.6	3.0	1.2-4.5	4.2	1.9-7.9
MAM	1.8	0.5-2.8	2.9	1.2-5.1	4.4	1.9-8.0
JJA	1.8	0.9-2.2	3.0	1.5-5.4	4.4	1.9-8.5
SON	1.3	0.4-2.3	2.8	1.2-5.0	4.1	1.8-8.8
EASTERN U.S.						
DJF	1.8	0.4-2.6	2.6	1.4-5.8	4.6	2.2-10.2
MAM	1.7	0.6-3.2	2.7	1.4-6.0	4.4	1.9-9.6
JJA	1.6	0.8-1.9	2.8	1.4-5.5	4.2	1.8-8.6
SON	1.5	0.6-2.3	2.8	1.4-5.4	4.0	1.8-9.0

Some of the studies analyze only electricity. Almost all show both an increase in electricity consumption and an increase in the consumption of primary fuels used to generate it, except in the few regions that provide space heating with electricity (for example, the Pacific Northwest). The few studies that examine effects on peak electricity demand emphasize that increases in peak demand would cause disproportionate increases in energy infrastructure investment.

Some studies provide demand estimates for heating fuels such as natural gas and distillate fuel oil in addition to electricity. These all-fuels studies provide support for the idea that climate warming causes significant decreases in space heating; however, whether energy savings in heating fuels offset increases in energy demand for cooling depends on the initial balance of energy consumption between heating and cooling, which in turn depends upon geography. Empirical studies show that the overall effect is more likely to be a significant net savings in delivered energy consumption in northern parts of the country (those with more than 4,000 heating degree-days per year) and a significant net increase in energy consumption in the south for both residential and commercial buildings, with the national balance slightly favoring net savings of delivered energy.

Studies vary in their treatment of the expected demographic shifts in the United States, expected evolution of building stock, and consumer reaction to warmer temperatures. Roughly half of the studies use building energy simulation models and account explicitly for the current trend in U.S. population moving toward the south and west, as well as increases in square footage per capita in newer buildings and increases in market penetration of air conditioning in newer buildings (See Annex A for a summary of methods). They do not, however, include consumer reactions to warming itself. For example, the market penetration of air conditioning is not directly influenced by warming in these studies. The other studies use econometric modeling of energy consumption choices. Many of these studies emphasize that the responsiveness of climate change of energy use to climate change is greater in the long-run than in short run; for example, consumers not only run their air conditioners more often in response to higher temperatures, but may also adopt air conditioning for the first time in regions such as New England, which still feature relatively low market penetration of air conditioning. Commercial building designs may evolve to reduce the need for heating by making better use of internal energy gains and warmer weather. Rising costs of space conditioning could modify the current trend in floor space per capita. Most econometric studies of building energy consumption estimate effects like this statistically from databases on existing buildings such as the Energy Information Administration's (EIA's) Residential Energy Consumption Survey (RECS) (EIA,

2001b) and Commercial Building Energy Consumption Survey (CBECS) (EIA, 2003).

When losses in energy conversion and delivery of electricity are taken into account, primary energy consumption (source energy) at the national level increases in some studies and decreases in others, with the balance of studies projecting a net increase in primary energy consumption. When the higher costs per delivered Btu of electricity are taken into account, the national-level consumer expenditures on energy increase in some studies and decrease in others, with the balance of studies favoring an increase in expenditures.

Various studies include a range of climate warming scenarios as well as different time frames and methods. Table 2.2 summarizes the main qualitative conclusions that can be drawn from an overview of this literature concerning the marginal effect of climate warming on energy use in buildings. These effects are discussed further in Sections 2.3 through 2.5.

Table 2.2. Summary of Qualitative Effects of Global Warming on Energy Consumption in the United States

Sector	National Effects	Regional Effects	Other Effects	Comments
Residential and Commercial Buildings Annual Energy Use	Slight decrease or increase in net annual delivered energy; likely net increase in primary energy	Space heating savings dominate in North; space cooling increases dominate in South	Overall increase in carbon emissions	Studies agree on the direction of regional effects; national direction varies with the study
Peak Electricity Consumption	Probable increase	Increase in summer peaking regions; probable decline in winter peaking regions	Increase in carbon emissions	Most regions are summer-peaking due to air conditionning
Market Penetation of Energy-Using Equipment	Increase in market penetration of air conditioning	Air conditioning market share increases primarily in North	—	Very few studies. Strength of the effect is not clear.

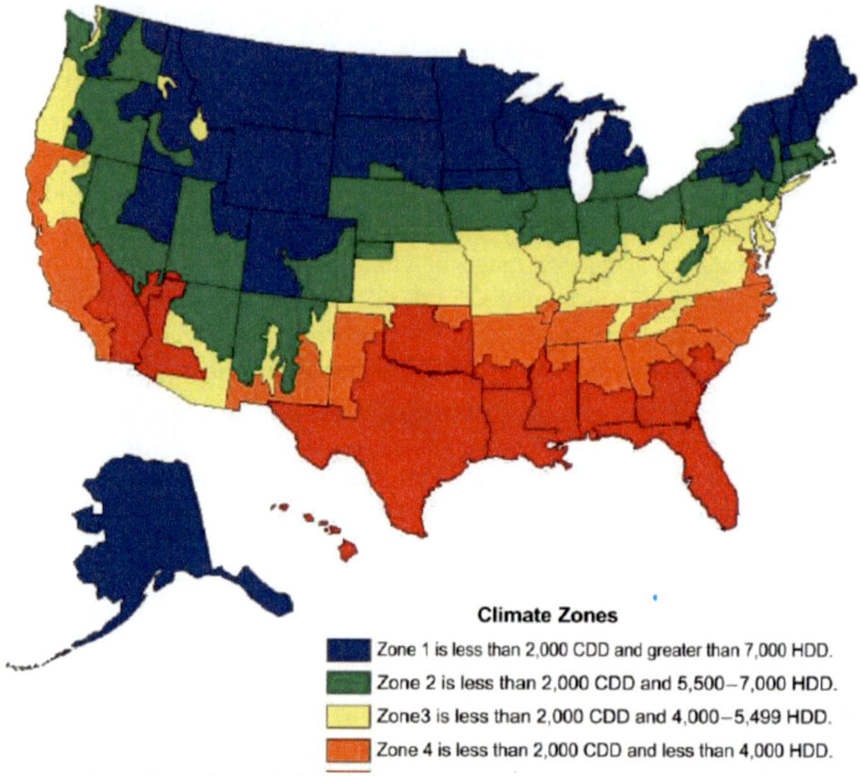

Source: Enery Information Administration, Residential Energy Consumption Survey (EIA, 2001c). *http://www.eia.doe.gov/*

Figure 2.1. U.S. Climate Zones (Zones 1-3 are "North," Zones 4-5 are "South").

2.2.2. The Literature in Greater Detail

The general finding about the net impact of climate warming on the consumption of delivered heating fuel and electricity is that for regions with more than about 4000 heating degree-days Fahrenheit (EIA Climate Zones 1-3, roughly the dividing line between "north" and "south" in most national studies—see Figure 2.1) climate warming tends to reduce consumption of heating fuel more than it increases the consumption of electricity (e.g., Hadley et al., 2004, 2006). The reverse is true south of that line. By coincidence, the national gains and losses in delivered energy approximately balance. Existing studies do not agree on whether there is small increase or decrease. The picture is different for primary energy and carbon dioxide. Because the generation, transmission, and distribution of electricity is subject to significant energy losses, national primary energy demand tends to increase with warmer temperatures. Finally, because electricity is about 50% generated with coal, which is a high-carbon fuel, and about 3.2 Btu of primary energy are consumed for every Btu of delivered electricity (EIA, 2006), carbon dioxide emissions also tend to increase. The extent of this national shift in energy use is expected to depend in part on the strength of residential adoption of air conditioning as the length of the air conditioning season and the warmth of summer increases in the north, where the market penetration of air conditioning is still relatively low. The potential reaction of consumers to a longer and more intense cooling season in the future has been addressed in only a handful of studies (e.g., Sailor and Pavlova, 2003) and must be considered highly uncertain. There is even less information available on theoff-setting effects of adaptations such as improved energy efficiency or changes in urban form that might reduce exacerbating factors such as urban heat island effects.

Box 2.1 provides insight into the recent trends in the intensity of energy consumption in residential and commercial buildings in the United States. There are a number of underlying trends, such as an ongoing population shift to the South and West, increases in the floor space per building occupant in both the residential and commercial sectors, and improvements in building shell performance, the balance of which have led to overall reductions in the intensity in the use of fuels for heating. Climate warming could be expected to reinforce this trend. At the same time, the demographic shifts to the South and West, increases in floor space per capita, and electrification of the residential and commercial sectors all have increased the use of electricity, especially for space cooling. This trend also would be reinforced by climate warming.

Amato et al., (2005) observe that many studies worldwide have analyzed the climate sensitivity of energy use in residential, commercial, and industrial buildings and have used these estimated relationships to explain energy consumption and to assist energy suppliers with short-term planning (Quayle and Diaz, 1979; Le Comte and Warren, 1981; Warren and LeDuc, 1981; Downton et al., 1988; Badri, 1992; Lehman, 1994; Lam, 1998; Yan, 1998; Morris, 1999; Considine, 2000; Pardo, et al., 2002). The number of studies in the U.S. analyzing the effects of climate *change* on energy demand, however, is much more limited. One of the very early national-level studies was of the electricity sector, projecting that between 2010 and 2055 climate change could increase capacity addition requirements by 14–23% relative to nonclimate change scenarios, requiring investments of $200–300 billion ($1990) (Linder and Inglis, 1989). The Linder-Inglis results are similar to electricity findings in most of the studies that followed. Subsequently, a number of studies have attempted an "all fuels" approach and have focused on whether net national energy demand (decreases in heating balanced against increases in cooling) would increase or decrease in residential and commercial buildings as a result of climate change (e.g., Loveland and Brown, 1990; Rosenthal et al., 1995; Belzer et al., 1996; Hadley et al., 2004, 2006; Mansur et al., 2005; Scott et al., 2005; Huang, 2006). The picture here is more clouded. While the direction of regional projections in these studies are reasonably similar, the net impacts at the national level differ among studies and depend on the relative balance of several effects, including scenarios used, assumptions about demographic trends and building stock, market penetration of equipment (especially air conditioning), and consumer behavior.

BOX 2.1 TRENDS IN THE ENERGY INTENSITY OF RESIDENTIAL AND COMMERCIAL BUILDINGS

Energy Use, Activity, Intensity and Other Factors in the Residential Sector – Delivered Energy, 1985-2004

Total energy use of delivered energy in households increased from 1985 to 2004. While both the number of households and housing size has increased over the period, the weather-adjusted intensity of energy use has fallen. Heating and cooling energy use declined, while appliance energy use increased enough to offset the declines in other end-uses. EIA (2006) projects an increase in building residential floor space per household of 14% during the period 2003-2030.

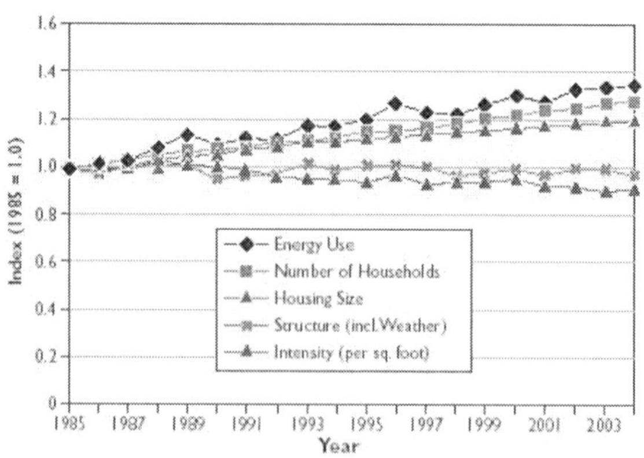

Commercial Energy Use, Activity, Weather, and Intensity - Delivered Energy

Estimated total floor space in commercial buildings grew 35% during the 1985-2004 period, while weather-adjusted energy intensity remained about constant. Declines in 1991 and since 2001 resulted from recessions, during which commercial vacancies increased and the utilization of occupied space fell. EIA (2006) projects the ratio of commercial floor space per member of the U.S. labor force to increase by 23% in the period 2003-2030.

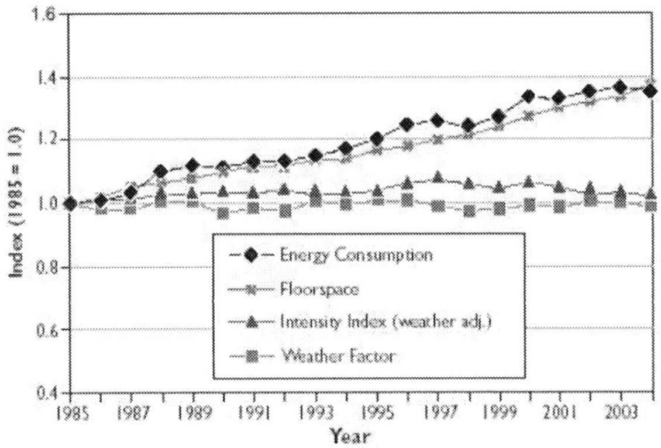

(Data from the U.S. Department of Energy, Office of Energy Efficiency and Renewable Energy, "Indicators of Energy Intensity in the United States," http://intensityindicators.pnl.gov/index.stm) and from EIA's Annual Energy Outlook (EIA, 2006).

In the sections that follow, this chapter discusses the impacts of climate warming on space heating in buildings (divided between residential and commercial), space cooling (again divided between residential and commercial buildings), net energy demand, market penetration of air conditioning, and possible effects of adaptation actions such as increased energy efficiency and changes to urban form, which could reduce the impacts of some compounding effects such as urban heat islands.

2.3. EFFECTS OF CLIMATE WARMING ON ENERGY USE FOR SPACE HEATING

2.3.1. Residential Space Heating

Temperature increases resulting from global warming are almost certain to reduce the amount of energy needed for space heating in residential buildings in the United States. The amount of the reduction projected by a number of U.S. studies has varied, depending mainly on the amount of temperature change in the climate scenario, the calculated sensitivity of the building stock to warming, and the adjustments allowed in the building stock over time (Table 2.3).

Table 2.3. Effects of Climate Change on Residential Space Heating in U.S. Energy Studies

Study: Author(s) and Date	Change in Energy Consumption (%)	Temperature Change (°C) and Date for Change
National Studies		
Rosenthal et al., 1995	-14%	+1°C (2010)
Scott et al., 2005	-4% to -20%	+About 1.7°C median (varies from 0.4° to 3.2°C regionally and seasonally) (2020)
Mansur et al., 2005	-2.8% for electricity-only customers; -2% for gas customers; -5.7% for fuel oil customers	+1° C January temperatures (2050)
Huang, 2006	Varies by location and building. vintage average HVAC changes: -12% heating in 2020 -24% heating in 2050 -34% heating in 2080	18 US locations, (varies by location, month, and time of day) Average winter temperature increases 1.3° C in 2020 2.6° C in 2050 4.1° C in 2080
Regional Studies		
Loveland and Brown, 1990	-44 to -73%	3.7°C to 4.7°C (Individual cities) (No date given)
Amato et al., 2005 (Massachusetts)	-7% to -14%, natural gas -15% to 20%, fuel oil -15% to -25%, natural gas -15% to -33%, fuel oil	-8.7% in HDD (2020) -11.5% in HDD (2030)
Ruth and Lin, 2006 (Maryland)	-2.5% natural gas -2.7% fuel oil	1.7°C-2.2°C (2025)

In most areas where it is available, the fuel of choice for residential and commercial space heating is natural gas, which is burned directly in a furnace in the building in question. There are some exceptions. In the Northeast, some of these savings will be in fuel oil, since fuel oil still provides about 36 % of residential space heating in that region, according to the 2001 RECS. In some other parts of the country with relatively short, mild winters or relatively inexpensive electricity or both, electricity has a significant share of the space heating market. For example, electricity accounted for 15% of residential heating energy in the Pacific Census Division and 19% in the South Atlantic Census Division in 2001 (EIA, 2001).

In Mansur et al., the impact of climate change on the consumption of energy in residential heating is relatively modest. When natural gas is available, the marginal impact of a 1°C increase in January temperatures in their model is predicted to reduce residential electricity consumption by 2.8% for electricity-only consumers and 2% for natural gas customers.

Table 2.4. Effects of Climate Change on Commercial Space Heating in U.S. Energy Studies

Study: Author(s) and Date	Change in Energy Consumption (%)	Temperature Change (°C) and Date for Change
National Studies		
Rosenthal et al., 1995	-16%	+1°C (2010)
Belzer et al., 1996	-29.0% to -35%	+3.9°C (2030)
Scott et al., 2005	-5% to -24%	+About 1.7°C median (varies from 0.4° to 3.2°C regionally and seasonally) (2020)
Mansur et al., 2005	-2.6% for electricity; -3% for natural gas; -11.8% for fuel oi	+1° C January temperatures (2050)
Huang, 2006	Varies by location and building vintage; Average heating savings: -12% heating in 2020; -22% heating in 2050; -33% heating in 2080	5 US locations, (varies by location, month, and time of day) Average winter temperature increases 1.3° C in 2020; 2.6° C in 2050; 4.1° C in 2080

Table 2.4. (Continued)

Study: Author(s) and Date	Change in Energy Consumption (%)	Temperature Change (°C) and Date for Change
Regional Studies		
Loveland and Brown, 1990	37.3% to -58.8% 3.7°C to	3.7°C to4.7°C (Individual cities) (No date given)
Scott et al., 1994 (Minneapolis and Phoenix)	-26.0% (Minneapolis); -43.1% (Phoenix)	3.9°C (no date)
Amato et al., 2005 (Massachusetts)	-7% to -8% -8% to 13%	-8.7% in HDD (2020) -11.5% in HDD (2030)
Ruth and Lin, 2006 (Maryland)	-2.7% natural gas	1.7°C-2.2°C (2025)

Scott et al., (2005), working directly with residential end uses in a building energy simulation model, projected about a 4% to 20% reduction in the demand for residential space heating energy by 2020, given no change in the housing stock and with winter temperature increases ranging from 0.4° to 3.2° C, or roughly 6% to 10% decrease in space heating per degree C increase. This is roughly twice the model sensitivity of Mansur et al., 2005. The Scott et al. analysis utilized the projected seasonal ranges of temperatures in Table 2.1 (Ruosteenoj a et al., 2003). Huang, 2006 also found decreases in average energy use for space heating. While these varied considerably by location and building vintage as well, the overall average was about a 12% average site energy reduction for space heating in 2020, or 9.2% per 1°C.

Regional level studies show similar effects, with a sensitivity of about 6% to 10% per 1°C in temperature change among the studies using building models and only about 1% per degree 1°C in studies using econometrics, in part possibly due to reactive increases in energy consumption (energy consumption "take-backs") as heating energy costs decline with warmer weather in this type of model, but also due to choice of region. In two studies with many of the same researchers and using very similar methodologies, Amato et al., 2005 projected about a 7% to 33% decline in space heating in the 2020s in Massachusetts, which has a long heating season, while Ruth and Lin, 2006 projected only a 2%-3% decline space heating energy during the same time frame in Maryland, which has a much milder heating season and many days where warmer weather would have no impact on heating degree-days or heating demand.

2.3.2. Commercial Space Heating

Although historically the intensity of energy consumption in the commercial sector has not followed a declining trend in the residential sector (Box 2.1), the effects of climate warming on space heating in the commercial sector (Table 2.4) are projected in most studies to be similar to those in the residential sector.

Belzer et al., (1996) used the detailed CBECS data set on U.S. commercial buildings, and calculated the effect of building characteristics and temperature on energy consumption in all U.S. temperature on energy consumption in all U.S. commercial buildings. With building equipment and shell efficiencies frozen at 1990 baseline levels and a 3.9°C temperature change, the Belzer model predicted a decrease in annual space heating energy requirements of 29% to 35%, or about 7.4% to 9.0% per 1°C. Mansur et al. 2005 projected that a 1°C increase in January temperatures would produce a reduction in electricity consumption of about 3% for electricity for all-electric customers. The warmer temperatures also would reduce natural gas consumption by 3% and fuel oil demand by a sizeable 12% per 1°C. This larger impact on fuel oil consumption likely occurs because warming has its largest impacts on heating degree days in the Northeast and in some other northern tier states where fuel oil is most prevalent. Another factor may be the fact that commercial buildings that use fuel oil may be older vintage buildings whose energy consumption is more sensitive to outdoor temperatures. Similar to its residential findings, Huang, 2006 showed that the impact of climate change on commercial building energy use varies greatly depending on climate and building type. For the entire U.S. commercial sector, the simulations showed 12% decrease in energy use for space heating or 9.2% per 1°C.

Again, the regional level studies produce more dramatic decreases in energy demand in colder regions than in warmer ones; however, the differences are less between cold regions and warm regions than in residential buildings because commercial buildings are more dominated by internal loads such as lighting and equipment than are residential buildings.

2.4. Effects of Climate Warming on Energy Use for Space Cooling

2.4.1. Residential Space Cooling

According to all studies surveyed for this chapter, climate warming is expected to significantly increase the energy demand in all regions for space cooling, which is provided almost entirely by electricity. The effect in most studies is nonlinear with respect to temperature and humidity, such that the percentage impact increases more than proportionately with increases in temperature (Sailor, 2001). Some researchers have projected that increases in cooling eventually could dominate decreases in heating as temperatures continue to rise (Rosenthal et al., 1995), although that effect is not necessarily observed in empirical studies for the temperature increases projected in the United States during the 21st century (Table 2.5).

Electricity demand for cooling was projected to increase by roughly 5% to 20% per 1ºC for the temperature increases in the national studies surveyed. This can differ by location and customer class. For example, Mansur et al., 2005 projected that when July temperatures were increased by 1ºC, electricity-only customers increased their electricity consumption by 4%, natural gas customers increased their demand for electricity by 6%, and fuel oil customers bought 15% more electricity. The impact on all electricity consumption is somewhat lower because electricity also is used for a variety of non-climate-sensitive loads in all regions and for space heating and water heating in some regions. Looking specifically at residential sector cooling demand (rather than all electricity) with a projected 2020 building stock, Scott et al. 2005 projected nationally that an increase of 0.4° to 3.2°C summer temperatures (Table 2.5) results in a corresponding 8% to 39% increase in national annual cooling energy consumption, or roughly a 12% to 20% increase per 1°C. Huang's (2006) projections show an even stronger increase of about a 38% increase in 2020 for a 1.7°C increase in temperature, or 22.4% per 1°C, perhaps in part because of differences in the in the details of locations and types of new buildings in particular, which tend to have more cooling load and less heating load.

Table 2.5. Effects of Climate Change on Residential Cooling Space in U.S. Energy Studies

Study: Author (s) and Date	Change in Energy Consumption (%)	Temperature Change (°C) and Date for Change Table 2.5. Effects of
National Studies		
Rosenthal et al., 1995	+20%	+1°C (2010)
Scott et al., 2005	+8% to +39%	+About 1.7°C median (varies from 0.4° to 3.2°C regionally and seasonally) (2020)
Mansur et al., 2005	4% for electricity only customers; 6% for natural gas customers; 15.3% for fuel oil customers	+1° C July temperatures (2050)
Huang, 2006	Varies by location and building vintage; Average HVAC savings: +38% heating in 2020 +89% heating in 2050 +158% heating in 2080	18 US locations, (varies by location, month, and time of day) Average summer temperature increases 1.7° C in 2020 3.4° C in 2050 5.3° C in 2080
Regional Studies		
Loveland and Brown, 1990	+55.7% to +146.7%	3.7°C to 4.7°C (Individual cities) (No date given)
Sailor, 2001	+0.9% (New York) to +11.6% (Florida) per capita	2°C (No date given)
Sailor and Pavlova, 2003 (Four states)	+13% to +29%	1°C (No date given)
Amato et al., 2005 (Massachusetts)	+6.8% in summer +10% to +40% (summer)	+12.1% in CDD (2020) +24.1% in CDD (2030)
Ruth and Lin, 2006 (Maryland)	+2.5% in May-Sep. (high energy prices); +24% (low energy prices)	1.7°C-2.2°C (2025)

Among the state studies, Loveland and Brown, 1990 found very high residential cooling sensitivities in a number of different locations across the country. Cooling energy consumption increased by 55.7% (Fort Worth, from a relatively high base) up to 146% (Seattle, from a very low base) for a temperature increase of 3.7°C to 4.7°C. This implies about a 17% to 31% increase in cooling energy consumption per degree C. Using a similar model in the special case of California, where space heating is already dominated by space cooling, Mendelsohn, 2003 projected that total energy expenditures for electricity used for space cooling would increase nonlinearly and that net overall energy expenditures would increase with warming in the range of 1.5°C, more for higher temperatures. In such mild cooling climates, relatively small increases in temperature can have a large impact on air-conditioning energy use by reducing the potentials for natural ventilation or night cooling. The residential electricity results in Sailor, 2001, Sailor and Pavlova, 2003; for several locations, and Amato et al., 2005 for Massachusetts are consistent with the national studies, with the expected direction of climate effects and about the expected magnitude, but the Ruth et al., 2006 results for the more southerly state of Maryland turn out to be very sensitive to electricity prices, ranging from +2.5% at high prices (about 8 cents per kWh, 1990$) prices to +24% if prices were low (about 6 cents per kWh, 1990$).

2.4.2. Commercial Space Cooling

U.S. studies also have projected a significant increase in energy demanded for space cooling in commercial buildings as a result of climate warming, as summarized in Table 2.6.

Commercial sector studies show that the percentage increases in space cooling energy consumption tend to be less sensitive to temperature than are the corresponding energy increases in the residential sector for the same temperature increase. For example, Rosenthal et al. 1995 found residential cooling increased 20% but commercial sector cooling only 15% for a 1°C temperature increase. The increase in Scott et al. 2005 had a range of 9.4% to 15% per 1°C for commercial and 12% to 20% per 1°C for residential customers. As with heating, in both cases this is likely to be in part because of the relatively greater sensitivity of space conditioning to internal loads in commercial buildings. Mansur et al., 2005 econometric results were less clear in this regard, possibly because geographic and behavioral differences among customer classes tend to obscure the overall effects of the buildings themselves. With building equipment and shell efficiencies frozen at 1990 baseline levels, Belzer et al., 1996 found impacts in the same range as the other studies. A 3.9°C temperature change increased annual space cooling energy requirements by 53.9% or about 9.0% to 13.8% per 1°C. Huang, 2006 also showed strong increases in cooling energy consumption at the national level. In

2020, his average increase was 17% for a 1.7°C temperature increase, or +10% per 1°C.

Table 2.6. Effects of Climate Change on Commercial Space Cooling in U.S. Energy Studies

Study: Author(s) and Date	Change in Energy Consumption (%)	Temperature Change (°C) and Date for Change	Comments
National Studies			
Rosenthal et al., 1995	+15%	+1°C (2010)	Energy weighted national averages of census division-level data
Belzer et al., 1996	+53.9%	+3.9°C (2030)	
Scott et al., 2005	+6% to +30%	+About 1.7°C median (varies from 0.4° to 3.2°C regionally and seasonally) (2020)	Varies by region
Mansur et al., 2005	+4.6% (electricity-only customers); -2% (natural gas customers); +13.8% (fuel oil customers)	+1° C January temperatures (2050)	A negative effect on electricity use for natural gas cus-tomers is statistic-cally significant at the 10% level, but unexplained
Huang, 2006	Varies by location and building vint-age; Average HVAC savings: +17% heating in 2020 +36% heating in 2050 +53% heating in 2080	5 US locations, (varies by location, month, and time of day) Average winter temperature increases 1.7° C in 2020 3.4° C in 2050 5.3° C in 2080	

Table 2.6. (Continued)

Study: Author(s) and Date	Change in Energy Consumption (%)	Temperature Change (°C) and Date for Change	Comments
Regional Studies			
Loveland and Brown, 1990 (general office buildings in 6 individual cities)	+34.9% in Chicago +75.0% in Seattle	3.7°C to 4.7°C (Individual cities) (No date given)	
Scott et al., 1994 (small office buildings in specific cities)	58.4% in Minneapolis 36.3% in Phoenix	3.9°C (no date)	
Sailor, 2001 (7 out of 8 energy-intensive states; one state–Washington–used electricity for space heating)	+1.6% in New York; +5.0% in Florida (per capita)	2°C (No date given)	
Amato et al., 2005 (Massachusetts)	+2% to +5% summer +4% to +10% summer	+12.1% in CDD (2020) +24.1% in CDD (2030)	Monthly per employee
Ruth and Lin, 2006 (Maryland)	+10% per employee in Apr-Oct	+2.2°C (2025)	

State-level studies generally show impacts that are in the same range as their national counter- parts. Analyses performed with building energy models generally indicate a 10% to 15% electric energy increase for cooling per 1°C. The econometric studies also show increases, but because the numerator is generally the change in consumption of all electricity (including lighting and plug loads, for example) rather than just that used for space cooling, the percentage increases are much smaller.

2.4.3. Other Considerations: Market Penetration of Air Conditioning and Heat Pumps (All-Electric Heating and Cooling), and Changes in Humidity

Although effects of air conditioning market penetration were not explicitly identified, the late-1990s econometrically based cross-sectional studies of Mendelsohn and colleagues might be interpreted as accounting for increased long run market saturations of air conditioning because warmer locations in the cross-sectional studies have higher market saturations of air conditioning as well as higher usage rates. However, more recent studies have examined the effects directly. In one example, Sailor and Pavlova, 2003 have projected that potential increases in market penetration of air conditioning in the residential sector in response to warming might have an effect on electricity consumption larger than the warming itself. They projected that although the temperature-induced increases in market penetration of air conditioning had little or no effect on residential energy consumption in cities such as Houston (93.6% market saturation), in cooler cities such as Buffalo (25.1% market saturation) and San Francisco (20.9% market saturation), the extra market penetration of air conditioning induced by a 20% increase in CDD more than doubled the energy use due to temperature alone. Using cross-sectional data and econometric techniques Mendelsohn, 2003 and Mansur et al., 2005 also have estimated the effects of the market penetration of space cooling into the energy market. Mansur et al. found that warmer winter temperatures were associated with higher likelihood of all-electric space conditioning systems in the sample survey of buildings in EIA's RECS and CBECS datasets. In warmer regions they noted that electricity has a high marginal cost but a low fixed cost, making it desirable in moderate winters. Electric heating is currently more prevalent in the South than in the North (EIA, 2001a). In general, however, the effects of adaptive market response of air conditioning to climate change have not been studied thoroughly in the United States.

High atmospheric humidity is known to have an adverse effect on the efficiency of cooling systems in buildings in the context of climate change because of the energy penalty associated with condensing water. This was demonstrated for a small commercial building modeled with the DOE-2 building energy simulation model in Scott et al., (1994), where the impact of an identical temperature increase created a much greater energy challenge for two relatively humid locations (Minneapolis and Shreveport), compared with two drier locations (Seattle and Phoenix). A humidity effect does not always show up in empirical studies (Belzer et al., 1996), but Mansur et al., 2005 modeled the effect of high humidity by introducing a rainfall as a proxy variable for humidity into their cross- sectional equations. In their residential sector, a one-inch increase in monthly precipitation resulted in more consumption by natural gas users of both electricity (7%) and of natural gas (2%). In their commercial sector, a one-inch increase in July precipitation resulted in more consumption of natural gas (6%) and of fuel oil (40%).

2.5. Overall Effects of Climate Change on Energy Use in Buildings

2.5.1. Annual Energy Consumption

Many of the U.S. studies of the impact of climate change on energy use in buildings deal with both heating and cooling and attempt to come to a "bottom line" net result for either total energy site consumed or total primary energy consumed (that is, both the amount of natural gas and fuel oil consumed directly in buildings and the amount of natural gas, fuel oil, and coal consumed indirectly to produce the electricity consumed in buildings.) Some studies only deal with total energy consumption or total electricity consumption and do not decompose end uses as has been done in this chapter. Recent studies show similar net effects. Both net delivered energy and net primary energy consumption increase or decrease only a few percent; however, there is a robust result that, in the absence of an energy efficiency policy directed at space cooling, climate change would cause a significant increase in the demand for electricity in the United States, which would require the building of additional electricity generation (and probably transmission facilities) worth many billions of dollars.

In much of the United States, annual energy used for space heating is far greater than space cooling; so net use of delivered energy would be reduced by global warming. Table 2.7 summarizes the results from a number of U.S. studies of the effects of climate change on net energy demand in U.S. residential and commercial buildings. The studies shown in Table 2.7 do not entirely agree with each other because of differences in methods, time frame, scenario, and geography. However, they are all broadly consistent with a finding that, at the national level, expected temperature increases through the first third of 21st Century (Table 2.1) would not significantly increase or decrease net energy use in buildings. The Linder and Inglis, 1989 projections concerning increases in electricity consumption have been generally confirmed by later studies, but there are geographical differences. For example, Sailor's state level econometric analyses (Sailor and Muñoz, 1997, Sailor, 2001, Sailor and Pavlova, 2003) projected a range of effects. A temperature increase of 2°C would be associated with an 11.6% increase in residential per capita electricity used in Florida (a summer-peaking state dominated by air conditioning demand), a 5% increase per 1°C warming. On the other hand, a 7.2% decrease in Washington state (which uses electricity extensively for heating and is a winter-peaking system), had about a 3% decrease per 1°C warming.

The Rosenthal et al., 1995 projections of reduced net total delivered energy consumption and energy expenditure reductions have not been confirmed. Results of more recent studies follow a temperature increase of 2°C that would be associated with an 11.6% increase in residential per capita electricity used in Florida (a summer-peaking state dominated by air conditioning demand) and a 5% increase per 1°C warming. On the other hand, a 7.2% decrease in Washington state (which uses electricity extensively for heating and is a winter-peaking system), had about a 3% decrease per 1°C warming.

Table 2.7. Climate Change Effects in Combined Residential-Commercial Studies and Combined Results from Sector Studies

Study: Author(s) and Date	Change in Energy Consumption (%)	Temperature Change (°C) and Date for Change	Comments
National Studies			
Linder-Inglis, 1989	+0.8% to +1.6% Annual electricity consumption; +3.4% to +5.1% annual elec-tricity consumption.	+0.8°C to +1.5°C (2010) +3.5°C to +5.0°C (2050)	Results available for 47 state and substate service areas
Rosenthal, et al., 1995	-11% Annual energy load; balance of hea-ting and cooling nationally.	1°C (2010)	Space heating and air condit-ioning combined
Mendelsohn, 2001	+1% to +22% Residential expenditures -11% to +47% Commercial Expenditures	+1.5°C to +5°C (2060)	Takes into account energy price fore-casts, market penetration of air conditioning. Precipitation increases 7%.

Table 2.7. (Continued)

Study: Author(s) and Date	Change in Energy Consumption (%)	Temperature Change (°C) and Date for Change	Comments
Scott et al., 2005	-2% to -7% (Reside-ntial and commercial heating and cooling consumption comb-ined (site energy). Energy used for cool-ing increases, heating energy decreases.	About +1.7°C median (varies from +0.4° to +3.2°C regionally and seasonally) (2020)	Varies by region. Allows for grow-th in residential and commercial building stock, but not increased adoption of air conditioning in response to warming
Mansur et al., 2005	+2% Residential expenditures, 0% commercial expenditures	+1°C Annual temperature (2050)	Takes into acco-unt energy price forecasts, market penetration of air conditioning. Precipitation increases 7%.
Hadley et al., 2004, 2006	Heating -6%, cooling +10%, +2% primary	+1.2°C (2025)	Primary energy, residential
	energy Heating -11% cooling +22% -1.5% primary energy	+3.4°C (2025)	and commercial combined. Allows for growth in residential and commercial building stock.
Huang. 2006	Varies by location, building type and vin-tage average HVAC changes: -8% site, +1% primary in 2020 -13% site, +0% prima-ry in 2050 -15% site, +4% primary in 2080	18 U.S. locations (varies by city, month, and time of day); average summer tempera-ture increases: 1.7° C in 2020 3.4° C in 2050 5.3° C in 2080	Huang. 2006
Loveland and Brown, 1990	+10% to +35% HVAC load in general offices; -22.0% to +48.1% HVAC load in single-family houses	+3.2°C to +4.0°C ($2xCO_2$, no date)	Multiple state study: results are for individual areas
Sailor, 2001 (8 energyin-tensive states; electricity only)	Residential: -7.2% in Washington to +11.6% in Florida Commercial: -0.3% (Washington) to +5% in Florida	+2°C (Derived from IPCC; but no date given)	

Scott et al., 2005 projected that overall site energy consumption in U.S. residential and commercial buildings is likely to decrease by about 2% to 7% in 2020 (0.4°C to 3.2°C warming). This amounts to about 2% per 1°C warming, which is in the same direction of the Rosenthal et al. results, but smaller. This effect takes into account expected changes in the building stock, but not increased market penetration of air conditioning that specifically results from climate change. For a 1°C increase in year-round temperatures, Mansur et al., 2005 provide only projections of net energy expenditures—a 2% increase in total residential energy expenditures -- and no net change in commercial energy demand for the year 2060. In residences, electricity expenditures (presumably mainly for cooling) generally increase, while use of other fuels generally decreases. Projected commercial sector expenditures show increases in electricity expenditures that are almost exactly offset by declines in the expenditures for natural gas and fuel oil. Since the Mansur et al. analysis claims to estimate long-term climate elasticities that include fuel choices and comfort choices as well as the direct effect of warmer temperatures on building energy loads, its results likely reflect at least some of the increased adoption of air conditioning that would be expected in residences in currently cooler climates as temperatures increase; residential sector electricity use is projected to grow faster than electricity use in the commercial sector, where air conditioning is more common and internal loads such as lighting dominate electricity use. Hadley et al., 2004, 2006 also project cooling energy consumption increasing and heating energy consumption decreasing. The projected national net effect on delivered energy consumption is slightly negative; but the impact on primary energy consumption is a slight increase. For all three studies, the impact of 1°C to 2°C warming is small. At the individual city level, Loveland and Brown, 1990 projected lower residential energy load in northern cities such as Chicago, Minneapolis, and Seattle and increased energy loads in southern cities such as Charleston, Ft. Worth, and Knoxville. A general office building increase showed increased overall energy loads in all six cities.

Most recently, Huang, 2006 used results from the HADCM3 GCM that project the changes in temperature, daily temperature range, cloud cover, and relative humidity by month for 0.5° grids of the earth's surface to produce future weather files for 18 U.S .locations. under 4 IPCC climate change scenarios (A1FI, A2M, B1, and B2M) for three time periods (2020, 2050, and 2080). These weather files were then used with the DOE-2 building energy simulation program to calculate the changes in space conditioning energy use for a large set of prototypical residential and commercial buildings to represent the U.S. building stock. This study looked in detail at the technical impact of climate change on space conditioning energy use, but did not address socio-economic factors or adaptive strategies to climate change.

These simulations indicate that the overall impact of climate change by 2020 on the U.S. building stock would be a 7% reduction in site energy use, corresponding to a 1% reduction in primary energy, when the generation and transmission losses for electricity are taken into account. The savings were noticeably larger for residential buildings (9% reduction in site and 2% reduction in primary energy use) than for commercial buildings (7% reduction in site, but a 3% increase in primary energy use). The counterbalancing effect of heating savings in the north, however, tends to mask the appreciable impact that climate change can have on cooling-dominant locations in the south. For example, cooling energy use in single-family houses in Miami and New Orleans was expected to increase by about 20%. In the North or West, the percentage increase of cooling was actually much larger, but due to the short cooling season, the savings were more than offset by the reductions in heating energy use. For example, cooling energy use was expected to rise by 100% in San Francisco, 60% in Boston and Chicago, and 50% in New York and Denver.

Because of their larger internal heat gains and less exposure to the outdoors in commercial buildings, these simulations project that commercial buildings would require less heating and more cooling than residential houses. Consequently, some building types such as large hotels and supermarkets showed an increase in site energy use with climate change, and almost all showed increases in primary energy use. In Los Angeles and Houston, commercial building energy use would increase by 2% and 4% in site energy use, and by 15% and 25% in primary energy use.

Huang, 2006 also looked at the impact of climate change out to 2050 and 2080, where there are cumulative effects of further temperature increases coupled with newer, tighter buildings that require much less heating and proportionally more cooling than older existing buildings. By 2050, heating loads were expected to be reduced by 28%, and cooling loads increased by 85% due to climate change, averaged across all building types and climates. By 2080, heating loads were expected to be reduced by nearly half (45%), but cooling loads were expected to more than double (165%) due to climate change, averaged across all building types and climates. With falling energy use for heating and rising energy use for cooling, by 2080 the ratio of cooling to heating energy use would be 60% in site energy and close to 180% in primary energy.

There are also a number of specific regional-level studies with similar outcomes. For Massachusetts in 2020, Amato et al., 2005 projected a 6.6% decline in annual heating fuel consumption (8.7% decrease in heating degree days— overall temperature change not given) and a 1.9% increase in summer electricity consumption (12% increase in annual cooling degree- days). Amato et al. noted that per capita residential and commercial energy demands in Massachusetts are sensitive to temperature and that a range of climate warming scenarios may noticeably decrease winter heating fuel and electricity demands and increase summer electricity demands. For 2030, the estimated residential summer monthly electricity demand projected increases averaged about 20% to 40%. Wintertime monthly natural gas demand declined by 10% to 20%. Fuel oil demand was down about 15% to 30%. For the commercial sector, electricity consumption rose about 6% to 10%. Winter natural gas demand declined by 6% to 14%.

The Hadley et al., 2006 study used the DD- NEMS energy model. Two advantages of this approach are that it provides a direct comparison at the regional level to official forecasts and that it provides a fairly complete picture of energy supply, demand, and endogenous price response in a market model. One disadvantage is that the DD-NEMS model only projected to 2025 at the time of that study (now 2030), which is only the earliest part of the period where climate change is expected to substantially affect energy demand. Hadley's regional results were broadly similar to those in Scott et al., 2005. For example, they showed decreases in energy demand for heating, more than offsetting the increased demand for cooling in the north (New England, Mid-Atlantic, West North Central and especially East North Central Census Division). In the rest of the country, the increase in cooling was projected to dominate. Nationally, the site energy savings were shown to be greater than the site energy increases, but because of energy losses in electricity generation, primary energy consumption (primary energy) increased by about 3% by 2025, driving up the demand for coal and driving down the demand for natural gas. Also, because electricity costs more than natural gas per delivered Btu, the increase in total energy cost per year was found to be about $15 billion (2001 dollars).

2.5.2. Peak Electricity Consumption

Studies published to date project that temperature increases with global warming would increase peak demand for electricity in most regions of the country. The amount of the increase in peak demand would vary with the region. Study findings vary with the region or regions covered and the study methodology— in particular, whether the study allows for changes in the building stock and increased market penetration of air conditioning in response to warmer conditions. The Pacific Northwest, which has significant market penetration of electric space heat, relatively low market penetration of air conditioning, and a winter-peaking electric system, is likely to be an exception to the general rule of increased peak demand. The Pacific Northwest power system annual and peak demand would likely be lower as a result of climate warming (Northwest Power and Conservation Council, 2005).

Concern for peak electricity demand begins with the earliest studies of the climate impacts on building energy demand. Linder and Inglis, 1989, in their multiregional study of regional electricity demand, found that although annual electricity consumption increased from +3.4 to +5.1% , peak electricity demand would increase between 8.6% and 13.8% , and capacity requirements between 13.1% and 19.7%, costing tens of billions of dollars.

One of the other few early studies of the effects of climate change on regional electricity was conducted by Baxter and Calandri, 1992. The case of California has received particular attention (See Box 2.2). For instance, the study used degree day changes from General Circulation Model (GCM) projections for 2010 to adjust the baseline heating and cooling energy uses in residential and commercial models that were derived from building energy simulations of prototypical buildings. Two climate change scenarios were considered; a low temperature increase scenario of 0.72°C in the winter, 0.60°C in the spring and fall, and 0.48°C in the summer, and a high temperature increase scenario of 2.28°C in the winter, 1.90°C in the spring and fall, and 1.58°C in the summer. Results were presented for the five major utility districts, and showed a 0.28% decrease in heating coupled with a 0.55% increase in cooling energy use for the low-temperature increase scenario, and a 0.85% decrease in heating coupled with a 2.54% increase in cooling energy use for the high-temperature increase scenario. The statewide impacts on energy demand were a 0.34-1.51% increase in cooling electricity demand for the low- temperature increase scenario, and a 2.57-2.99% increase in cooling electricity demand for the high-temperature increase scenario.

The authors concluded that the impacts of climate change appear moderate on a percentage basis, but because California's electricity system is so large, a moderate percentage increase results in sizeable absolute impacts. For energy use, the 0.6% and 2.6% increases for the two scenarios signify increases of 1741 GWh and 7516 GWh. For electricity demand, the 0.34- 1.51% and 2.57-2.99% increases correspond to increased peak demand by 221-967 MW and 1648-1916 MW. To put these impacts in perspective, uncertainties in the state's economic growth rate would have had comparable or larger impacts on electricity demand over this 20-year projected estimation. Actual growth in noncoincident peak demand between 1990 and 2004 was actually 8,650 MW for total end use load and 9,375 MW for gross generation (California Energy Commission, 2006).

Much more recently, using IPCC scenarios of climate change from the Hadley3, PCM, and GFDL climate models downscaled for California, Franco and Sanstad 2006 found a high correlation between the simple average daily temperature and daily peak electricity demand in the California Independent System Operator region, which comprises most of California. They evaluated three different periods: 2005- 2034, 203 5-2064, and 2070-2099. In the first period, depending on the scenario and model, peak summer demand was projected to increase relative to a 1961-1990 base period before climate change by 1.0%-4.8%; in the second, 2.2%-10.9%; in the third, 5.6%-19.5%.

BOX 2.2 CALIFORNIA'S PERSPECTIVE ON CLIMATE CHANGE

There has been probably more analysis done in California on impacts of climate change than anywhere else in the U.S. (also see Box 5.1). The reasons for this are: (1) California's relative mild climate has been shown to be highly sensitive to climate change, not only in terms of temperature, but also in water resources, vegetation distribution, and coastal effects, and (2) California is vulnerable to shortfalls in peak electricity demand, as demonstrated by the electricity shortage in 2001 (albeit mostly man-made) and the recent record heat wave in July 2006 that covered the entire state and was of greater intensity and longer duration than previously recorded. The pioneering work by Baxter and Calandri, 1992 on global warming and electricity demand in California has already been described elsewhere in this report (see main text, this section). Mendelsohn, 2003 investigated the impact of climate change on energy expenditures, while Franco 2005, Franco and Sanstad 2006, and Miller et al,. 2006 have all focused on the impact of climate change on electricity demand. Miller et al., 2006 studied the probability of extreme weather phenomena under climate change scenarios for California and other Western U.S. locations. GCMs show that, over time, California heat waves will have earlier onsets, be more numerous, and increase in duration and intensity. "For example, extreme heat days in Los Angeles may increase from 12 to as many as 96 days per year by the end of the century, implying current-day heat wave conditions may extend the entire summer period". Overall, projected increases in extreme heat by 2070-2099 will approximately double the historical number of days for inland California cities, and up to four times for coastal California cities like Los Angeles and San Diego. The following plots show how the duration of extreme periods in California increases based on GCM results (from Miller et al., 2007).

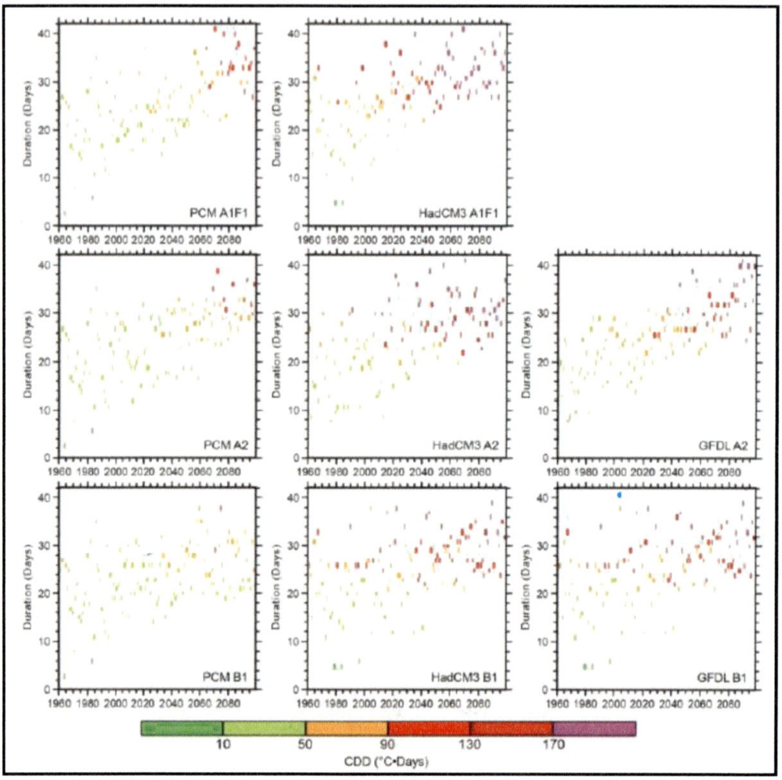

A few U.S. regions could benefit from lower winter demand for energy in Canada. An example is in the New England-Middle Atlantic- East North Central region of the country, where Ontario and Québec in particular are intertied with the U.S. system, and where demand on either side of the international border could influence the other side. For example, since much of the space heating in Québec is provided by hydro-generated electricity, a decline in energy demand in the province could free up a certain amount of capacity for bordering U.S. regions in the winter. In Québec, the Ouranos organization (Ouranos, 2004) has projected that net energy demand for heating and air conditioning across all sectors could fall by 30 trillion Btus, or 9.4 % of 2001 levels by 2100. Seasonality of demand also would change markedly. Residential heating in Québec would fall by 15%

and air conditioning (currently a small source of demand) would increase nearly fourfold. Commercial-institutional heating demand was projected to fall by 13% and commercial air conditioning demand to double. Peak (winter) electricity demand in Québec would decline. Unfortunately, Québec's summer increase in air conditioning demand would coincide with an increase of about 7% to 17% in the New York metropolitan region (Ouranos, 2004); so winter savings might be only of limited assistance in the summer cooling season, unless the water not used for hydroelectric production in the winter could be stored until summer and the transmission capacity existed to move the power south (Québec's hydroelectric generating capacity is sized for the winter peak and should not be a constraint).

Although they discuss the impacts of climate change on peak electricity demand, Scott et al., 2005 did not directly compute them. However, they performed a sensitivity analysis using nuclear power's 90% average capacity factor for 2004 as an upper-thatbound estimate of base load power plant availability and projected that national climate sensitive demand consumption (1.3 quads per year by 2080) would be equivalent of roughly 48 GW, or 48 base load power plants of 1,000 MW each. At the much lower 2003 average U.S. generation/capacity ratio of 47%, 93 GW of additional generation capacity would be required. This component of demand would be a factor in addition to any increases due to additional climate-related market penetration of air conditioning and any other causes of increased demand for electricity that the national electrical system will be dealing with for the rest of the century.

For further information about methods for estimating energy consumption in buildings, see Annex A.

2.6. Adaptation: Increased Efficiency and Urban Form

Although improving building energy efficiency should help the nation cope with impacts of climate change, there is relatively little specific information available on the potential impacts of such improvements. Partly this is because it has been thought that warming would already be reducing energy consumption, so that the additional effects of energy efficiency have not been of much interest. Scott et al., 1994 and Belzer et al., 1996 concluded that in the commercial sector, very advanced building designs could increase the savings in heating energy due to climate warming alone. Loveland and Brown 1990, Scott et al. 1994, Belzer et al., 1996, and Scott et al., 2005 all estimated the effects of energy-efficient buildings on energy consumption in the context of climate change and also concluded that much of the increase in cooling energy consumption due to warming could be offset by increased energy efficiency.

Loveland and Brown, 1990 projected that changes leading to -50% lighting, +50% insulation, and +75% window shading would reduce total energy use in residential buildings by 3 1.5% to 44.4% in the context of a 3.2° to 4°C warming. This suggests that advanced building designs are a promising approach to reducing energy consumption impacts of warming, but further verification and follow-up research is needed both to confirm results and design strategies.

Scott et al., 1994 examined the impact of "advanced" building designs for a 48,000-square foot office building in the context of climate change in the DOE-2 building energy simulation model. The building envelope was assumed to reduce heat transfer by about 70% compared to the ASHRAE 90.1 standard. It included extra insulation in the walls and ceiling, reduction in window conductivity by a factor of 6, and window shading devices. The result was that, assuming a 3.9°C increase in annual average temperature, rather than experiencing between an 8% reduction in energy use (Minneapolis) and a 6.3% increase in overall energy use (Phoenix), an advanced design building would show a 57.2% to 59.8% decrease in energy used. In addition, the cooling energy impact was reversed in sign–a 47% to 60% decrease instead of a 35% to 93% increase. Cost, however, was not analyzed (also see SAP 4.6).

Belzer et al., 1996 projected that with a 3.9°C increase in annual average temperature, the use of advanced buildings would increase the overall energy savings in EIA's year 2030 projected commercial building stock from 0.47 quads (20.4%) to 0.63 quads (27%). Use of advanced building designs in the 2030 commercial building stock would increase the overall energy savings by 1.15 quads (40.6%) relative to a 2030 building stock frozen at 1990 efficiency. The cooling component of building energy consumption was only reduced rather than reversed by advanced designs in this study.

Finally, Scott et al., 2005 explicitly considered the savings that might be achieved under the Department of Energy's energy efficiency programs as projected in August 2004 for the EIA building stock in the year 2020 (temperature changes of about 0.4°C at the low end to about 2.8°C at the high end). This is the only study to have estimated the national effects of actual energy efficiency programs in the context of global warming. (The analysis did not count any potential increase in energy demand due to additional climate change-induced market penetration of air conditioning). The efficiency programs, which mainly targeted heating, lighting, and appliances instead of cooling, were less effective if the climate did not change; however, buildings still saved between 2.0 and 2.2

quads. This was a savings of about 4.5%, which would more than offset the growth in temperature-sensitive energy consumption due to increases in cooling and growth in building stock between 2005 and 2020.

Except for Scott et al., 2005, even where studies consider adaptive response (e.g., Loveland and Brown, 1990; Belzer, et al. 1996; Mendelsohn, 2001), they generally do not involve particular combinations of technologies to offset the effects of future climate warming. Regionally, Franco and Sanstad, 2006 did note that the very aggressive energy efficiency and demand response targets for California's investor-owned utilities such as those recently enacted by the California Public Utilities Commission could, if extended beyond the current 2013 horizon -- provide substantial "cushioning" of the electric power system against the effects of higher temperatures.

2.7. Other Possible Effects, Including Energy Use in Key Sectors

2.7.1. Industry

Except for energy used to heat and cool buildings, which is thought to be about 6% of energy use in industry (EIA, 2001b) and is generally not analyzed for manufacturing activities in existing studies, it is not thought that industrial energy demand is particularly sensitive to climate change. For example, Amato et al. 2005 stated that "industrial energy demand is not estimated since previous investigations (Elkhafif, 1996; Sailor and Munoz, 1997) and our own findings indicate that it is non-temperature-sensitive." Ruth and Lin, 2006 observe that in contrast to residential households, which use about 58% of their energy for space conditioning, and commercial buildings, which use about 40%, industrial facilities devote only about 6% of their energy use to space conditioning. In absolute numbers, this is about a third of what the commercial sector uses and about 8% of what the residential sector uses for this purpose. According to the 2002 Manufacturing Energy Consumption Survey, among the energy uses that could be climate sensitive, U.S. manufacturing uses about 4% of all energy for directly space conditioning, 22% for process heating, and 1.5% for process cooling (EIA, 2002a).

This does not mean, of course, that industry is not sensitive to climate, or even that energy availability as influenced by climate or weather does not affect industry. Much of the energy used in industry is used for water heating; so energy use would likely decline in industry if climate and water temperatures become warmer. Electrical outages (some caused by extreme weather) cause many billions in business interruptions every year, and large events that interrupt energy supplies are also nationally important (see Chapter 3). However, little information exists on the impact of climate change on energy use in industry. Considine, 2000 econometrically investigated industrial energy use data from the EIA Short Term Energy Forecasting System based on HDD and CDD and calculated that U.S. energy consumption per unit of industrial production would increase for an increase 0.0127% per increase in one heating degree day (Fahrenheit) or by 0.0032% per increase of one cooling degree day (Fahrenheit). On an annual basis with a 1°C temperature increase (1.8°F), there would be a maximum of 657 fewer HDD per year and 657 more CDD (Fahrenheit basis, and assuming that all industry was located in climates that experienced all of the potential HDD decrease and CDD increase). This would translate into 6.2% less net energy demand in industry or a saving of roughly 0.04 quads.

A few studies have focused on a handful of exceptions where it was assumed that energy consumption would be sensitive to warmer temperatures, such as agricultural crop drying and irrigation pumping (e.g., Darmstadter, 1993; Scott et al., 1993). While it seems logical that warmer weather or extended warm seasons should result in warmer water inlet temperatures for industrial processes and higher rates of evaporation, possibly requiring additional industrial water diversions, as well as additional municipal uses for lawns and gardens, the literature review conducted for this chapter did not locate any literature either laying out that logic or calculating any associated increases in energy consumption for water pumping. Industrial pumping increases are likely to be small relative to those in agriculture, which consumes the lion's share (40%) of all fresh water withdrawals in the United States (USGS, 2004). Some observations on energy use in other climate-sensitive economic sectors follow.

2.7.2. Transportation

Running the air conditioning in a car reduces its fuel efficiency by approximately 12% at highway speeds (Parker, 2005). A more extended hot season likely would increase the use of automotive air conditioning units, but by how much and with what consequences for fuel economy is not known. Based on preliminary unpublished data, virtually all new light duty vehicles sold (well over 99% in 2005) in the United Sates come with factory-installed air conditioning (up from about 90% in the mid-1990s)[1], but no statistics appear to be available from public sources on the overall numbers or percentage of vehicles in the fleet without air conditioning. No projections appear to be available on the total impact of climate change on energy consumption in automotive air conditioners; however, there are some estimates of the response of vehicle air conditioning use to temperature. Based on a modeling of consumer comfort, Johnson (2002) estimates that at ambient temperatures above 30°C (86°F), drivers would have their air conditioning on 100% of the time; at 21°C-30°C (70°F-86°F), 80%; at 13°C-20°C (55°F-70°F), 45%; and at 6°C-12°C (43°F-55°F), 20% of the time.[2] Data from the Environmental Protection Agency's model of vehicular air conditioning operation suggests that U.S. drivers on average currently have their air conditioning systems turned on 23.9% of the time. With an increase in ambient air temperature of 1°C (1.8°F), the model estimates that drivers would have their air conditioning systems turned on 26.9% of the time, an increase of 3.0% of the time.[3]

Much of the food consumed in the United States moves by refrigerated truck or rail. One of the most common methods is via a refrigerated truck-trailer combination. As of the year 2000, there were approximately 225,000 refrigerated trailers registered in the United States, and their Trailer Refrigeration Units (TRUs) used on average 0.7 to 0.9 gallons of fuel per hour to maintain 0°F. On a typical use cycle of 7200 hours per year (6 days per week, 50 weeks per year), the typical TRU would use 5,000 to 6,000 gallons of diesel per year (Shurepower, LLC, 2005), or between 26 and 32 million barrels for the national fleet. Even though diesel electric hybrid and other methods are making market inroads and over time could replace a substantial amount of this diesel use with electricity from the grid when the units are parked, climate warming would add to the energy use in these systems. No data appear to be available on the total impact of climate change on energy consumption in transportation, however (also see SAP 4.7).

2.7.3. Construction

Warming the climate should result in more days when outdoor construction activities are possible. In many parts of the northern states, the construction industry takes advantage of the best construction weather to conduct activities such as some excavation, pouring concrete, framing buildings, roofing, and painting, while sometimes enclosing buildings, partially heating them with portable space heaters, and conducting inside finishing work during "bad" weather. While the construction season may lengthen in the North, there also may be an increasing number of high-temperature heat stress days during which outdoor work may be hindered. The net effects on energy consumption on construction are not clear. The literature survey conducted for this chapter was not able to locate any studies in the United States that have investigated either the lengthening of the construction season in response to global warming or any resulting impacts on energy consumption.

2.7.4. Agriculture

Agricultural energy use generally falls into five main categories: equipment operations, irrigation pumping, embodied energy in fertilizers and chemicals, product transport, and drying and processing. A warmer climate implies increases in the demand for water in irrigated agriculture and use of energy (either natural gas or electricity) for pumping. Though not a factor in many parts of the country, irrigation energy is a significant source of energy demand west of the 100th meridian, especially in the Pacific Southwest and Pacific Northwest. For example, irrigation load in one early climate change impact assessment increased from about 8.7% to about 9.8% of all Pacific Northwest electricity load in July (Scott et al., 1993), even with no change in acreage irrigated.

In some parts of the country, the current practice is to keep livestock and poultry inside for parts of the year, either because it is too cold or too hot outside. Often these facilities are space-conditioned. In Georgia, for example, there are 11,000 poultry houses, and many of the existing houses are air-conditioned due to the hot summer climate (and all new ones are) (University of Georgia and Fort Valley State University, 2005). Poultry producers throughout the South also depend on natural gas and propane as sources of heat to keep their birds warm during the winter (Subcommittee on Conservation, Credit, Rural Development, and Research, 2001). The demand for cooling livestock and poultry would be expected to increase in a warmer climate, while that for heating of cattle barns and chicken houses likely would fall. There are no available quantitative estimates of the effects on energy demand.

Food processing needs extensive refrigerated storage, which may take more energy in a warmer climate. However, there seem to be no U.S. studies on this subject.

2.8. Summary of Knowledge about Possible Effects

Generally speaking, the net effects of climate change in the United States on total energy demand are projected to be modest, amounting to between perhaps a 5% increase and decrease in demand per 1°C in warming in buildings, about 1.1 Quads in 2020 based on EIA 2006 projections (EIA, 2006). Existing studies do not agree on whether there would be a net increase or decrease in energy consumption with changed climate because a variety of methodologies have been used. There are differences in climate sensitivities among models and studies as well as differences in methodological emphasis. For example, econometric models have incorporated some market response to warming and fuel costs but not necessarily differences in building size and technology over time and space, while the opposite is true of building simulation approaches. There are also differences in climate and market scenarios. It appears likely that some of the largest effects of climate change on energy demand are in residential and commercial buildings, however, with other sensitivities in other sectors being of secondary or tertiary importance.

Another robust finding is that most regions of the country can be expected to see significant increases in the demand for electricity, due both to increases in the use of existing space-cooling equipment and also to likely increases in the market penetration of air conditioning in response to longer and hotter summers. This is likely in Northern regions where market penetration of air conditioning is still relatively low.

To some extent, it is possible to control for differences in climate scenarios by comparing percentage changes in energy use per a standardized amount of temperature change, as has been done in this chapter. It is also possible to search for a set of robust results and to compare impacts, for example, that come from models that have fixed technologies and no market responses with those that allow technology to evolve and businesses and individuals to respond to higher or lower energy bills.

Some of the apparently conflicting results are more likely to be correct than others. Because of compensating market and technological responses, impacts of climate change should be less with models that allow technology to evolve and businesses and individuals to respond to higher or lower energy bills. Because they also assess more realistically the factors actually likely to be in play, they are likelier to be closer to correct. None of the models actually does all of this, but Mansur et al., 2005 probably comes the closest on the market side and Scott et al., 2005 or Huang, 2006 on the technology side. Using the results from these two approaches, together with Sailor and Pavlova, 2003 to inform and modify the

Hadley et al., 2006 special version of NEMS, probably has the best chance of being correct for buildings.

End Notes

[1] Data supplied by Robert Boundy, Oak Ridge National Laboratory, based on Ward's Automotive Yearbooks.

[2] Data supplied by Lawrence Chaney, National Renewable Energy Laboratory

[3] Data supplied by Richard Rykowski, Assessment Standards and Support Division, Environmental Protection Agency. The model used in this analysis is described in Chapter III of the Draft Technical Support Document to the proposed EPA rulemaking to devise EPA's methodology for calculating the city and highway fuel economy values pasted on new vehicles.

In: Effects of Climate Change on Energy Production… ISBN: 978-1-60741-426-1
Editor: Alrik M. Solberg

Chapter 4

EFFECTS OF CLIMATE CHANGE ON ENERGY PRODUCTION AND DISTRIBUTION IN THE UNITED STATES

U.S. Climate Change Science Program

Energy production in the U.S. is dominated by fossil fuels: coal, petroleum, and natural gas (Figure 3.!). Every existing source of energy in the United States has some vulnerability to climate variability (Table 3.!). Renewable energy sources tend to be more sensitive to climate variables; but fossil energy production can also be adversely effected by air and water temperatures, and the thermoelectric cooling process that is critical to maintaining high electrical generation efficiencies also applies to nuclear energy. In addition, extreme weather events have adverse effects on energy production, distribution, and fuel transportation.

This chapter discusses impacts on energy production and distribution in the United States associated with projected changes in temperature, precipitation, water resources, severe weather events, and sea level rise, although the currently available research literatures tend to be limited in most cases. Overall, the effects on the existing infrastructure might be categorized as modest; however, local and industry-specific impacts could be large, especially in areas that may be prone to disproportional warming (Alaska) or weather disruptions (Gulf Coast and Gulf of Mexico). The existing assemblage of power plants and distribution systems is likely to be more affected by ongoing unidirectional changes, compared with possible future systems, if future systems can be designed with the upfront flexibility to accommodate the span of potential impacts. Possible adaptation measures include technologies that minimize the impact of increases in ambient temperatures on power plant equipment, technologies that conserve water use for power plant cooling processes, planning at the local and regional level to anticipate storm and drought impacts, improved forecasting of the impacts of global warming on renewable energy sources at regional and local levels, and establishing action plans and policies that conserve both energy and water.

3.1. Effects on Fossil and Nuclear Energy

Climate change can affect fossil and nuclear energy production, conversion, and end-user delivery in a myriad of ways. Average ambient temperatures impact the supply response to changes in heating and cooling demand by affecting generation cycle efficiency, along with cooling water requirements in the electrical sector, water requirements for energy production and refining, and Gulf of Mexico (GOM) produced water discharge requirements. Often these impacts appear "small" based on the change in system efficiency or the potential reduction in reliability, but the scale of the energy industry is vast: fossil fuel-based net electricity generation exceeded 2,500 billion kWh in 2004 (EIA 2006). A net reduction in generation of 1% due to increased ambient temperature (Maulbetsch and DiFilippo 2006) would represent a drop in supply of 25 billion kWh that might need to be replaced somehow. The GOM temperature-related issue is a result of the formation of water temperature-related anoxic zones and is important because that region accounts for 20 to 30% of the total domestic oil and gas production in the U.S. (Figure 3.2). Constraints on produced water discharges could increase costs and reduce production, both in the GOM region and elsewhere. Impacts of extreme weather events could range from localized railroad track distortions due to temperature extremes, to regional-scale coastal flooding from hurricanes, to watershed-scale river flow excursions from weather variations superimposed upon, or possibly augmented by, climate change. Spatial scale can range from kilometers to continent-scale; temporal scale can range from hours to multiyear. Energy impacts of episodic events can linger for months or years, as illustrated by the continuing loss of oil and gas production in the GOM (MMS 2006a, 2006b, and 2006c) eight months after the 2005 hurricanes.

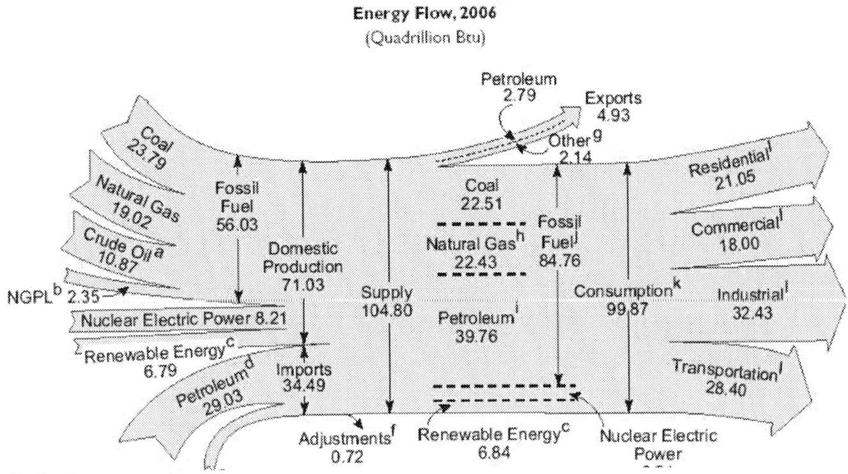

[a] Includes lease condensate.

[b] Natural gas plant liquids.

[c] Conventional hydroelectric power, biomass, geothermal, solar/PV, and wind.

[d] Crude oil and petroleum products. Includes imports into the Strategic Petroleum Reserve.

[e] Natural gas, coal, coal coke, fuel ethanol, and electricity.

[f] Stock changes, losses, gains, miscellaneous blending components, and unaccounted-for supply.

[g] Coal, natural gas, coal coke, and electricity.

[h] Natural gas only; excludes supplemental gaseous fuels.

[i] Petroleum products, including natural gas plant liquids and crude oil burned as fuel.

[j] Includes 0.06 quadrillion Btu of coal coke net imports.

[k] Includes 0.06 quadrillion Btu of electricity net imports.

[l] Primary consumption, electricity retail sales, and electrical systems energy losses, which are allocated to the end-use sectors in proportion to each sector's share of total electricity retail sales.

Sources: Tables 1.1, 1.2, 1.3, 1.4, 2.1a, and 10.1.

Notes: •Data are preliminary. •Values are derived from the source data prior to rounding for publication. •Totals may not equal sum of components due to independent rounding.

Figure 3.1. Energy Flow in the U.S. (EIA, *Annual Energy Review 2006*)

3.1.1. Thermoelectric Power Generation

Climate change impacts on electricity generation at fossil and nuclear power plants are likely to be similar. The most direct climate impacts are related to power plant cooling and water availability.

Projected changes in water availability throughout the world would directly affect the availability of water to existing power plants. While there is uncertainty in the nature and amount of the change in water availability in specific locations, there is agreement among climate models that there will be a redistribution of water, as well as changes in the availability by season. As currently designed, power plants require significant amounts of water, and they will be vulnerable to fluctuations in water supply. Regional-scale changes would likely mean that some areas would see significant increases in water availability, while other regions would see significant decreases. In those areas seeing a decline, the impact on power plant availability or even siting of new capacity could be significant. Plant designs are flexible and new technologies for water reuse, heat rejection, and use of alternative water sources are being developed; but, at present, some impact—significant on a local level—can be foreseen. An example of such a potential local effect is provided in Box 3.1— Chattanooga: A Case Study, which shows how cooling conditions might evolve over the 21st century for generation in one locality. Situations where the development of new power plants is being slowed down or halted due to inadequate cooling water are becoming more frequent throughout the U.S. (SNL, 2006b).

In those areas seeing an increase in stream flows and rainfall, impacts on groundwater levels and on seasonal flooding could have a different set of impacts. For existing plants, these impacts could include increased costs to manage on-site drainage and run-off, changes in coal handling due to increased moisture content or additional energy requirements for coal drying, etc. The following excerpt details the magnitude of the intersection between energy production and water use.

An October 2005 report produced by the National Energy Technology Laboratory stated, in part, that the production of energy from fossil fuels (coal, oil, and natural gas) is inextricably linked to the availability of adequate and sustainable supplies of water. While providing the United States with a majority of its annual energy needs, fossil fuels also place a high demand on the Nation's water resources in terms of both use and quality impacts (EIA, 2005d). Thermoelectric generation is water intensive; on average, each kWh of electricity generated via the steam cycle requires approximately 25 gallons of water, a weighted average that captures total thermoelectric water withdrawals and generation for both once-through and recirculating cooling systems. According to the United States Geological Survey (USGS), power plants rank only slightly behind irrigation in terms of freshwater withdrawals in the United States (USGS, 2004), although irrigation withdrawals tend to be more consumptive. Water is also required in the mining, processing, and transportation of coal to generate electricity all of which can have direct impacts on water quality. Surface and underground coal mining can result in acidic, metal-laden water that must be treated before it can be discharged to nearby rivers and streams. In addition, the USGS estimates that in 2000 the mining industry withdrew approximately 2 billion gallons per day of freshwater. Although not directly related to water quality, about 10% of total U.S. coal shipments were delivered by barge in 2003 (USGS, 2004). Consequently, low river flows can create shortfalls in coal inventories at power plants.

Freshwater availability is also a critical limiting factor in economic development and sustainability, which directly impacts electric-power supply. A 2003 study conducted by the Government Accountability Office indicates that 36 states anticipate water shortages in the next 10 years under normal water conditions, and 46 states expect water shortages under drought conditions (GAO 2003). Water supply and demand estimates by the Electric Power Research Institute (EPRI) for the years 1995 and 2025 also indicate a high likelihood of local and regional water shortages in the United States (EPRI 2003). The area that is expected to face the most serious water constraints is the arid southwestern United States.

Box 3.1 Chattanooga: A Case Study of Cooling Effects

A preliminary analysis of one IPCC climate change scenario (A1B) provides one example of how cooling conditions might evolve over the 21st century for generation in the Chattanooga vicinity (ORNL work in progress). In this example, a slight upward trend in stream flow would provide a marginal benefit for once- through cooling, but would be offset by increasing summertime air temperatures that trigger limits on cooling water intake and downstream mixed temperatures. Closed-cycle cooling would also become less effective as ambient temperature and humidity increased. Utilities would need to maintain generation capacity by upgrading existing cooling systems or shifting generation to newer facilities with more cooling capacity. Without technology-based improvements in cooling system energy efficiency or steam-cycle efficiency, overall thermoelectric generation efficiency would decrease.

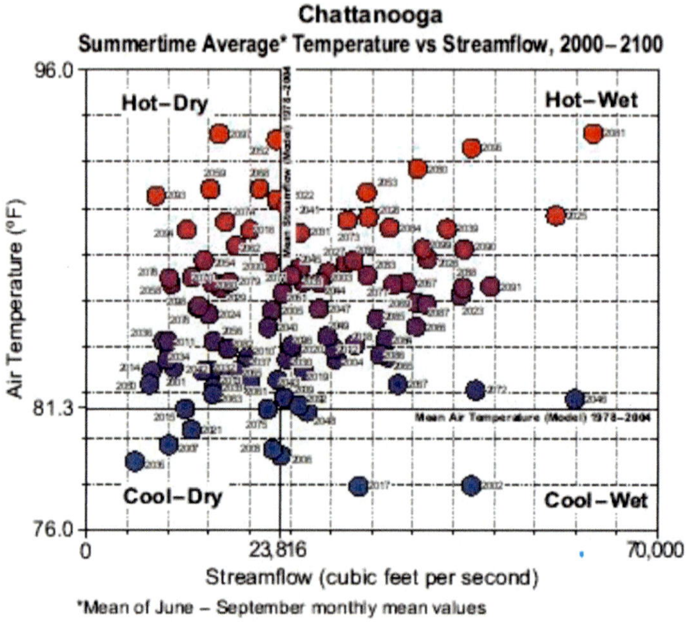

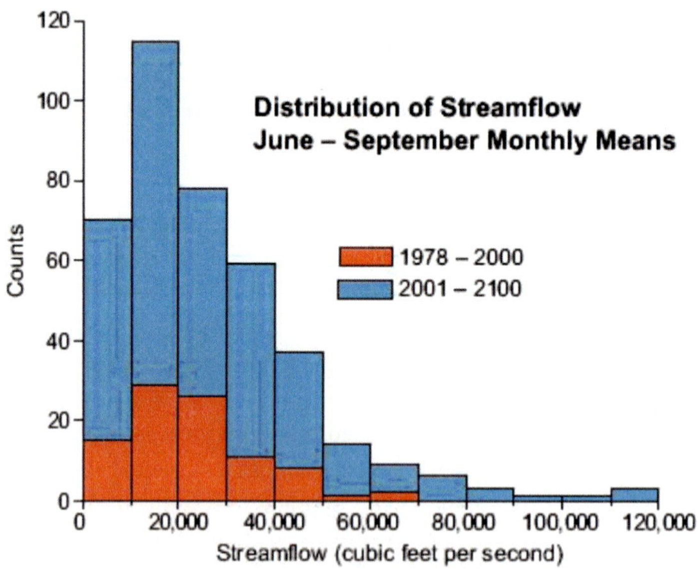

Table 3.1. Mechanisms of Climate Impacts on Various Energy Supplies in The U.S. Percentages Shown Are of Total Domestic Consumption

	Energy Impact Supplies	Climate Impact Mechanisms
Fossil Fuels (86%)	Coal (22%)	Cooling water quantity and quality (T), cooling efficiency (T, W, H), erosion in surface mining
	Natural Gas (23%)	Cooling water quantity and quality (T), cooling efficiency (T, W, H), disruptions of off-shore extraction (E)
	Petroleum (40%)	Cooling water quantity and quality, cooling efficiency (T, W, H), disruptions of off shore extraction and transport (E)
	Liquified Natural Gas (1%)	Disruptions of import operations (E)
Nuclear (8%)		Cooling water quantity and quality (T), cooling efficiency (T, W, H)
Renewables (6%)	Hydropower	Water availability and quality, temperature-related stresses, operational modification from extreme weather (floods/droughts), (T, E)
	Biomass	
	• Wood and forest products	Possible short-term impacts from timber kills or long-term impacts from timber kills and changes in tree growth rates (T, P, H, E, carbon dioxide levels)
	• Waste (municipal solid waste, landfill gas, etc.)	n/a
	• Agricultural resources (including derived biofuels)	Changes in food crop residue and dedicated energy crop growth rates (T, P, E, H, carbon dioxide levels)
	Wind	Wind resource changes (intensity and duration), damage from extreme weather
	Solar	Insolation changes (clouds), damage from extreme weather
	Geothermal	Cooling efficiency for air-cooled geothermal (T)

(T = water/air temperature, W = wind, H = humidity, P = precipitation, and E = extreme weather events)

Source: EIA, 2004

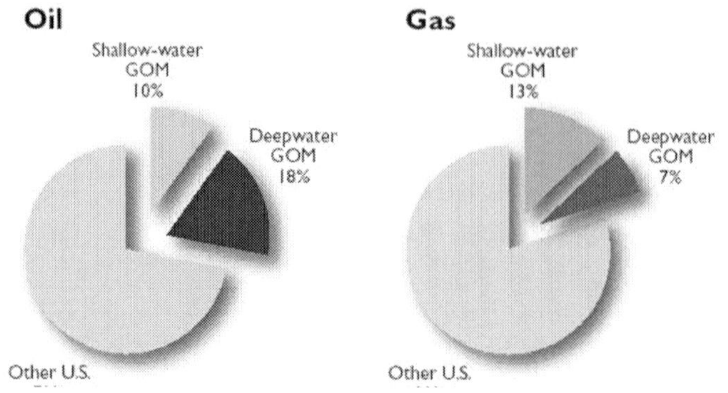

Figure 3.2. Distribution of Off-Shore Oil and Gas Wells in the Gulf of Mexico (GOM) and Elsewhere in the U.S.

In any event, the demand for water for thermoelectric generation will increasingly compete with demands from other sectors of the economy such as agriculture, residential, commercial, industrial, mining, and in-stream use. EPRI projects a potential for future constraints on thermoelectric power in 2025 for Arizona, Utah, Texas, Louisiana, Georgia, Alabama, Florida, and all of the Pacific Coast states. Competition over water in the western United States, including water needed for power plants, led to a 2003 Department of Interior initiative to predict, prevent, and alleviate water-supply conflicts (DOI 2003). Other areas of the United States are also susceptible to freshwater shortages as a result of drought conditions, growing populations, and increasing demand.

Concerns about water supply expressed by state regulators, local decision-makers, and the general public are already impacting power projects across the United States. For example, Arizona recently rejected permitting for a proposed power plant because of concerns about how much water it would withdraw from a local aquifer (Land Letter 2004). An existing Entergy plant located in New York is being required to install a closed-cycle cooling water system to prevent fish deaths resulting from operation of its once-through cooling water system (Greenwire, 2003). Water availability has also been identified by several Southern States Energy Board member states as a key factor in the permitting process for new merchant power plants (Clean Air Task Force 2004). In early 2005, Governor Mike Rounds of South Dakota called for a summit to discuss drought-induced low flows on the Missouri River and the impacts on irrigation, drinking-water systems, and power plants (Billingsgazette.com 2005). Residents of Washoe County, Nevada expressed opposition to a proposed coal-fired power plant in light of concerns about how much water the plant would use (Reno-Gazette Journal. 2005). Another coal-fired power plant to be built in Wisconsin on Lake Michigan has been under attack from environmental groups because of potential effects of the facility's cooling-water-intake structures on the Lake's aquatic life (Milwaukee Journal Sentinel, 2005).

Such events point toward a likely future of increased conflicts and competition for the water the power industry will need to operate their thermoelectric generation capacity. These conflicts will be national in scope, but regionally driven. It is likely that power plants in the west will be confronted with issues related to water rights: that is, who owns the water and the impacts of chronic and sporadic drought. In the east, current and future environmental requirements, such as the Clean Water Act's intake structure regulation, could be the most significant impediment to securing sufficient water, although local drought conditions can also impact water availability. If changing climatic conditions affect historical patterns of precipitation, this may further complicate operations of existing plants, and the design and site selection of new units.

EIA 2004a reports net summer and winter capacity for existing generating capacity by fuel source. Coal-fired and nuclear plants have summer/winter ratios of 0.99 and 0.98 and average plant sizes of 220 MW and 1015 MW, respectively. Petroleum, natural gas, and dual fuel-fired plants show summer/winter net capacity ratios of 0.90 to 0.93, indicating higher sensitivity to ambient temperature. Average sizes of these plants ranged from 12 MW to 84 MW, consistent with their being largely peaking and intermediate load units. Although large coal and nuclear generating plants report little degradation of net generating capacity from winter to summer conditions, there are reports (University of Missouri-Columbia 2004) of plant derating and shutdowns caused by temperature-related river water level changes and thermal limits on water discharges. Actual generation in 2004 (EIA, 2004a) shows coal-fired units with 32% of installed capacity provided 49.8% of generation and nuclear units with 10% of installed capacity provided 17.8% of power generated, indicating that these sources are much more heavily dispatched than are petroleum, natural gas, and dual-fired sources. To date, this difference has been generally attributed to the lower variable costs of coal and nuclear generation, indicating that the lower average dispatch has been more driven by fuel costs than temperature-related capacity constraints.

Gas turbines, in their varied configurations, provide about 20% of the electric power produced in the U.S. (EIA 2006). Gas turbines in natural gas simple cycle, combined cycle (gas and steam turbine), and coal-based integrated gasification combined cycle applications are affected by local ambient conditions, largely local ambient temperature and pressure. Ambient temperature and pressure have an immediate impact on gas turbine performance. Turbine performance is measured in terms of heat rate (efficiency) and power output. Davcock et al. (Davcock, DesJardins, and Fennell 2004) found that a 60°F increase in ambient temperature, as might be experienced daily in a desert environment, would have a 1-2 percentage point reduction in efficiency and a 20-25% reduction in power output. This effect is nearly linear; so a 10 degree Fahrenheit increase in ambient temperature would produce as much as a 0.5 percentage point reduction in efficiency and a 3-4% reduction in power output in an existing gas turbine. Therefore, the impact of potential climate change on the fleet of existing turbines would be driven by the impact that small changes in overall performance would have on both the total capacity available at any time and the actual cost of electricity.

Turbines for NGCC and IGCC facilities are designed to run 24 hours, 7 days a week; but simple cycle turbines used in topping and intermediate service are designed for frequent startups and rapid ramp rates to accommodate grid dispatch requirements. Local ambient temperature conditions will normally vary by 10 – 20°F on a 24-hour cycle, and many temperate-zone areas have winter-summer swings in average ambient temperature of 25-35°F. Consequently, any long-term climate change that would impact ambient temperature is believed to be on a scale within the design envelope of currently deployed turbines. As noted earlier, both turbine power output and efficiency vary with ambient temperature deviation from the design point. The primary impacts of longer periods of off-design operation will be modestly reduced capacity and reduced efficiency. Currently turbine-based power plants are deployed around the world in a wide variety of ambient conditions and applications, indicating that new installations can be designed to address long-term changes in operating conditions. In response to the range of operating temperatures and pressures to which gas turbines are being subjected, turbine designers have developed a host of tools for dealing with daily and local ambient conditions. These tools include inlet guide vanes, inlet air fogging (essentially cooling and mass flow addition), inlet air filters, and compressor blade washing techniques (to deal with salt and dust deposited on compressor blades). Such tools could also be deployed to address changes in ambient conditions brought about by longterm climate change.

3.1.2. Energy Resource Production and Delivery

Other than for renewable energy sources, energy resource production and delivery systems are mainly vulnerable to effects of sea level rise and extreme weather events.

IPCC 2001a projected a 50-cm. (20-in.) rise in sea level around North America in the next century from climate change alone. This is well within the normal tidal range and would not have any significant effect on off-shore oil and gas activities. On-shore oil and gas activities could be much more impacted, which could create derivative impacts on off-shore activities.

A number of operational power plants are sited at elevations of 3 ft or less, making them vulnerable to these rising sea levels. In addition, low-lying coastal regions are being considered for the siting of new plants due to the obvious advantages in delivering fuel and other necessary feedstocks. Significant percentages of other energy infrastructure assets are located in these same areas, including a number of the nation's oil refineries as well as most coal import/export facilities and liquefied natural gas terminals. Given that a large percentage of the nation's energy infrastructure lies along the coast, rising sea levels could lead to direct losses such as equipment damage from flooding or erosion or indirect effects such as the costs of raising vulnerable assets to higher levels or building future energy projects further inland, thus increasing transportation costs.

IPCC 2001 a and USGS 2000 have identified substantial areas of the U.S. East Coast and Gulf Coast as being vulnerable to sea-level rise. Roughly one-third of U.S. refining and gas processing physical plant lies on coastal plains adjacent to the Gulf of Mexico (GOM), hence it is vulnerable to inundation, shoreline erosion, and storm surges. On-shore but noncoastal oil and gas production and processing activities may be impacted by climate change primarily as it impacts extreme weather events, phenomena not presently well understood. Florida's energy infrastructure may be particularly susceptible to sea-level rise impacts. (See Box 3.2 Florida).

Alaska represents a special case for climate adaptation because the scale of projected impacts is expected to be greater in higher latitudes (See Box 3.3: A Case Study). Extreme weather events, which could represent more significant effects, are discussed in 3.1.4. Even coal production is susceptible to extreme weather events that can directly impact opencast mining operations and coal cleaning operations of underground mines.

Potential impacts on novel energy resources are speculative at present. Oil shale resource development, which is considered to be water intensive, could be made more difficult if climate change further reduces annual precipitation in an already arid region that is home to the major oil shale deposits. Water availability (Struck 2006) is beginning to be seen as a potential constraint on synthetic petroleum production from the Canadian oil sands. Coal-to-liquids operations also require significant quantities of water.

BOX 3.2 FLORIDA

Florida's energy infrastructure may be particularly susceptible to sea-level rise impacts. Most of the petroleum products consumed in Florida are delivered by barge to three ports (NASEO, 2005) two on the East Coast of Florida and one on the West Coast. The interdependencies of natural gas distribution, transportation fuel distribution and delivery, and electrical generation and distribution were found to be major issues in Florida's recovery from multiple hurricanes in 2004. In addition, major installations such as nuclear power plants are located very close to the seacoast at elevations very close to sea level. The map on the left shows major power plants susceptible to sea-level rise in Florida. The map on the right illustrates power plants in the path of Tropical Storm Ernesto.

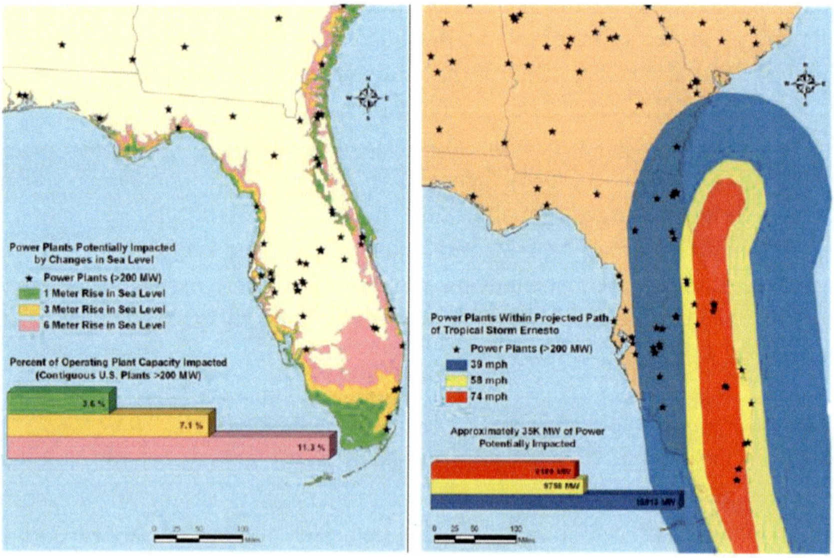

3.1.3. Transportation of Fuels

Roughly 65% of the petroleum products supplied in the Petroleum Administration for Defense (PAD) East Coast District (Figure 3.3) arrive via pipeline, barge, or ocean vessel (EIA 2004). Approximately 80% of the domestic-origin product is transported by pipeline. Certain areas, e.g., Florida, are nearly totally dependent on maritime (barge) transport. About 97% of the crude oil charged to PAD I refineries is imported, arriving primarily by ocean vessels. PAD II receives the bulk of its crude oil via pipeline, roughly two-thirds from PAD III and one-third from Canada. Both pipeline and barge transport have been susceptible to extreme weather events, with pipeline outages mostly driven by interdependencies with the electrical grid. In addition (see 3.3.2), increased ambient temperatures can degrade pipeline system performance, particularly when tied to enhanced oil recovery and, if practiced in the future, carbon sequestration. The transportation of coal to end users, primarily electrical generation facilities, is dependent on rail and barge transportation modes (EIA 2004b). Barge transport is susceptible to both short term, transient weather events and to longer-term shifts in regional precipitation and snow melt patterns that may reduce the extent of navigability of rivers and reduce or expand the annual navigable periods. In addition, offshore pipelines were impacted by Hurricane Ivan even before the arrival of Hurricanes Katrina and Rita (see 3.1.4).

Box 3.3 Alaska: A Case Study

Alaska represents a special case for climate adaptation where temperatures have risen (3°C) over the last few decades, a rate that is almost twice that of the rest of the world. Some models predict this warming trend will continue, with temperatures possibly rising as much as 4-7°C over the next 100 years (ACIA 2004).

In areas of Alaska's North Slope, change is already being observed. The number of days allowed for winter tundra travel dropped significantly since the state began to set the tundra opening date in 1969, and a chart of that decline has been widely used to illustrate one effect of a warming Arctic (Alaska Department of Natural Resources 2004). There is a significant economic impact on oil and natural gas exploration from a shorter tundra travel season, especially since exploration targets have moved farther away from the developed Prudhoe Bay infrastructure, requiring more time for ice road building. It is unlikely that the oil industry can implement successful exploration and development plans with a winter work season consistently less than 120 d.

Further, melting permafrost can cause subsidence of the soil, thereby threatening the structural integrity of infrastructure built upon it. It was anticipated that the Trans-Alaska Pipeline System would melt surrounding permafrost in the areas where it would be buried. Therefore, extensive soil sampling was conducted and in areas where permafrost soils were determined to be thaw-stable, conventional pipeline building techniques were utilized. But in ice-rich soils, the ground is generally not stable after the permafrost melts. Therefore, unique aboveground designs integrating thermal siphons were used to remove heat transferred in the permafrost via the pilings used to support the pipeline. And in a few selected areas where aboveground construction was not feasible, the ground around the pipeline is artificially chilled (U.S. Arctic Research Commission 2003 and Pipeline Engineering 2007). Such extensive soil testing and unique building techniques add substantial cost to large development projects undertaken in arctic climates but are necessary to ensure the long-term viability of the infrastructure.

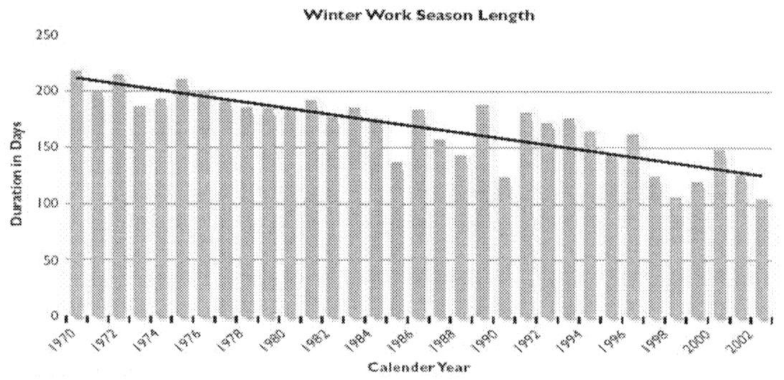

Source: ACIA, 2004.

Winter Work Season Length

Exploration in the Arctic may benefit from thinning sea ice. Recent studies indicate extent of sea ice covering the Arctic Ocean may have reduced as much as 10%, and thinned by as much as 15%, over the past few decades. These trends suggest improved shipping accessibility around the margins of the Arctic Basin with major implications for the delivery of goods as well as products such as LNG and oil from high latitude basins (ACIA 2004) . A reduction in sea ice may also mean increased off-shore oil exploration (ACIA 2004).

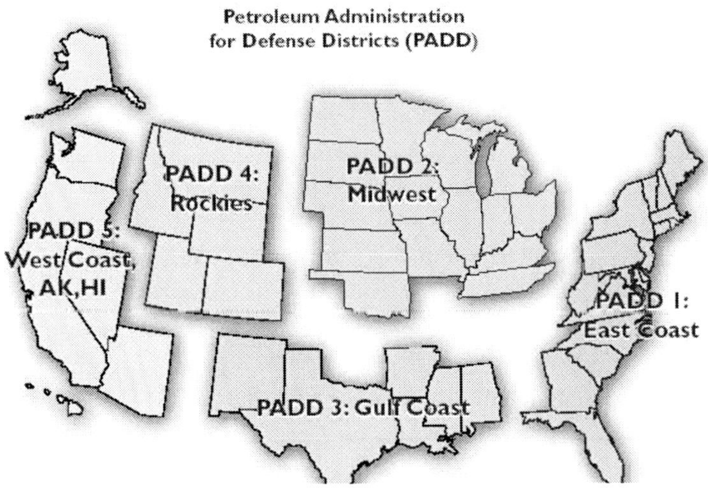

Figure 3.3. Petroleum Administration for Defense (PAD) Districts

3.1.4. Extreme Events

Climate change may cause significant shifts in current weather patterns and increase the severity and possibly the frequency of major storms (NRC 2002). As witnessed in 2005, hurricanes can have a debilitating impact on energy infrastructure. Direct losses to the energy industry in 2005 are estimated at $15 billion (Marketwatch.com 2006), with millions more in restoration and recovery costs. Future energy projects located in storm prone areas will face increased capital costs of hardening their assets due to both legislative and insurance pressures. For example, the Yscloskey Gas Processing Plant was forced to close for 6 months following Hurricane Katrina, resulting in both lost revenues to the plant's owners and higher prices to consumers as alternative gas sources had to be procured. In general, the incapacitation of energy infrastructure – especially of refineries, gas processing plants and petroleum product terminals – is widely credited with driving a price spike in fuel prices across the country, which then in turn has national consequences. The potential impacts of more severe weather are not, in fact, limited to hurricane-prone areas. Rail transportation lines, which transport approximately 2/3 of the coal to the nation's power plants (EIA 2002), often closely follow riverbeds, especially in the Appalachian region. More severe rainstorms can lead to flooding of rivers that then can wash out or degrade the nearby roadbeds. Flooding may also disrupt the operation of inland waterways, the second-most important method of transporting coal. With utilities carrying smaller stockpiles and projections showing a growing reliance on coal for a majority of the nation's electricity production, any significant disruption to the transportation network has serious implications for the overall reliability of the grid as a whole.

Off-shore production is particularly susceptible to extreme weather events. Hurricane Ivan (2004) destroyed seven GOM platforms, significantly damaged 24 platforms, and damaged 102 pipelines (MMS 2006). Hurricanes Katrina and Rita in 2005 destroyed more than 100 platforms and damaged 558 pipelines (MMS 2006). The two photographs in Figure 3.4 show the Mars deepwater platforms before and after the 2005 hurricanes. The $250 million Typhoon platform was so severely damaged that Chevron is working with the MMS to sink it as part of an artificial reef program in the GOM; the billion dollar plus Mars platform has been repaired and returned to production about 8 months post hurricane.

Figure 3.4. Hurricane damage in the Gulf of Mexico – Mars Platform

3.1.5. Adaptation to Extreme Events

Energy assets can be protected from these impacts both by protecting the facility or relocating it to safer areas. Hardening could include reinforcements to walls and roofs, the building of dikes to contain flooding, or structural improvements to transmission assets. However, the high cost of relocating or protecting energy infrastructure drives many companies to hedge these costs against potential repair costs if a disaster does strike. For example, it is currently estimated to cost up to $10 billion to build a new refinery from the ground up (Petroleum Institute for Continuing Education undated), compared with costs to fully harden a typical at-risk facility against a hurricane and with the few million dollars in repairs that may or may not be required if a hurricane does strike. Relocation of rail lines also faces a similar dilemma. BNSF's capacity additions in the Powder River Basin are expected to cost over $200 million dollars to add new track in a relatively flat region with low land prices; changes to rail lines in the Appalachian region would be many times more due to the difficult topography and higher land acquisition costs.

Industry, government agencies, and the American Petroleum Institute met jointly in March 2006 (API 2006a) to plan for future extreme weather events. Interim guidelines for jackup (shallow water) rigs (API 2006b) and for floating rigs (API 2006c) have been developed. MMS, DOT, and several industry participants have formed a Joint Industry Program (JIP) (Stress Subsea, Inc. 2005) to develop advanced capabilities to repair damaged undersea pipelines.

3.2. Effects on Renewable Energy Production

Renewable energy production accounted for about 6% of the total energy production in the United States in 2005 (Figure 3.5); biomass and hydropower are the most significant contributors (EIA 2005d), and the use of renewable energy is increasing rapidly in other sectors such as wind and solar. Biomass energy is primarily used for industrial process heating, with substantially increasing use for transportation fuels and additional use for electricity generation. Hydropower is primarily used for generating electricity, providing 270 billion kWh in 2005 (EIA, 2005d). Wind power is the fastest growing renewable energy technology, with total generation increasing to 14 billion kWh in 2005 (EIA 2006). Because renewable energy depends directly on ambient natural resources such as hydrological resources, wind patterns and intensity, and solar radiation, it is likely to be more sensitive to climate variability than fossil or nuclear energy systems that rely on geological stores. Renewable energy systems are also vulnerable to damage from extreme weather events. At the same time, increasing renewable energy production is a primary means for reducing energy-related greenhouse gas emissions and thereby mitigating the impacts of potential climate change. Renewable energy sources are therefore connected with climate change in very complex ways: their use can affect the magnitude of climate change, while the magnitude of climate change can affect their prospects for use.

3.2.1. Hydroelectric Power

Hydropower is the largest renewable source of electricity in the United States. In the period 2000-2004, hydropower produced approximately 75% of the electricity from all renewable sources (EIA 2005d). In addition to being a major source of base-load electricity in some regions of the United States (e.g., Pacific Northwest states), hydropower plays an important role in stabilizing electrical transmission grids, meeting peak loads and regional reserve requirements for generation, and providing other ancillary electrical energy benefits that are not available from other renewables when storage is unavailable. Hydropower project design and operation is very diverse; projects vary from storage projects with large, multipurpose reservoirs to small run-of-river projects that have little or no active water storage. Approximately half of the U.S. hydropower capacity is federally owned and operated (e.g., Corps of Engineers, Bureau of Reclamation, and the Tennessee Valley Authority); the other half is at nonfederal projects that are regulated by the Federal Energy Regulatory Commission. Nonfederal hydropower projects outnumber federal projects by more than 10:1.

The interannual variability of hydropower generation in the United States is very high, especially relative to other energy sources (Figure 3.6). The difference between the most recent high (2003) and low (2001) generation years is 59 billion kWh, approximately equal to the total electricity from biomass sources and much more than the generation from all other non-hydropower renewables (EIA 2006). The amount of water available for hydroelectric power varies greatly from year to year, depending upon weather patterns and local hydrology, as well as on competing water uses, such as flood control, water supply, recreation, and instream flow requirements (e.g., conveyance to downstream water rights, navigation, and protection of fish and wildlife). The annual variability in hydropower is usually attributed to climate variability, but there are also important impacts from multiple use operational policies and regulatory compliance.

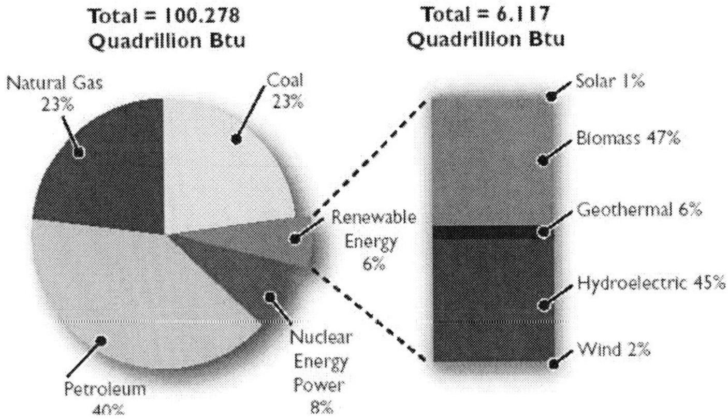

Figure 3.5. Renewable Energy's Share In U.S. Energy Supply (2005) (http://www.eia.doe. gov/cneaf/solar.rene wables/page/trens/hi ghlight 1 .html)

Figure 3.6. Historical Variability of Total Annual Production of Hydroelectricity From Conventional Projects in the U.S. (data from EIA Annual Energy Outlook, 2005).

There have been a large number of published studies on the climate impacts on water resource management and hydropower production (e.g., Miller and Brock 1988; Lettenmaier et al. 1999; Barnett et al. 2004). Significant changes are being detected now in the flow regimes of many western rivers (Dettinger 2005) that are consistent with the predicted effects of global warming. The sensitivity of hydroelectric generation to both changes in precipitation and river discharge is high, in the range 1.0 and greater (e.g., sensitivity of 1.0 means 1% change in precipitation results in 1% change in generation). For example, Nash and Gleick (1993) estimated sensitivities up to 3.0 between hydropower generation and stream flow in the Colorado Basin (i.e., change in generation three times the change in stream flow). Such magnifying sensitivities, greater than 1.0, occur because water flows through multiple power plants in a river basin. Climate impacts on hydropower occur when either the total amount or the timing of runoff is altered, for example when natural water storage in snow pack and glaciers is reduced under hotter climates (e.g., melting of glaciers in Alaska and the Rocky Mountains of the U.S.). Projections that climate change is likely to reduce snow pack and associated runoff in the U.S. West are a matter of particular concern.

Hydropower operations are also affected indirectly when air temperatures, humidity, or wind patterns are affected by changes in climate, and these driving variables cause changes in water quality and reservoir dynamics. For example, warmer air temperatures and a more stagnant atmosphere cause more intense stratification of reservoirs behind dams and a depletion of dissolved oxygen in hypolimnetic waters (Meyer et al. 1999). Where hydropower dams have tail-waters supporting cold-water fisheries for trout or salmon, warming of reservoir releases may have unacceptable consequences and require changes in project operation that reduce power production.

Evaporation of water from the surface of reservoirs is another important part of the water cycle that may be will be affected by climate change and may lead to reduced water for hydropower. However, the effects of climate change on evaporation rates is not straight-forward. While evaporation generally increases with increased air or water temperatures, evaporation also depends on other meteorological conditions, such as advection rates, humidity, and solar radiation. For example, Ohmura and Wild (2002) described how observed evapora evaporation rates decreased between 1950 and 1990, contrary to expectations associated with higher temperatures. Their explanation for the decrease was decreased solar radiation. Large reservoirs with large surface area, located in arid, sunny parts of the U.S., such as Lake Mead on the lower Colorado River (Westenburg et al., 2006), are the most likely places where evaporation will be greater under future climates and water availability will be less for all uses, including hydropower.

Competition for available water resources is another mechanism for indirect impacts of climate change on hydropower. These impacts can have far-reaching consequences through the energy and economic sectors, as happened in the 2000-2001 energy crises in California (Sweeney, 2002).

Recent stochastic modeling advances in California and elsewhere are showing how hydropower systems may be able to adapt to climate variability by reexamining management policies (Vicuña et al., 2006). The ability of river basins to adapt is proportional to the total active storage in surface water reservoirs (e.g., Aspen Environmental Group and M-Cubed, 200 5). Adaptation to potential future climate variability has both near-term and long-term benefits in stabilizing water supplies and energy production (e.g., Georgakakos et al., 2005), but water management institutions are generally slow to take action on such opportunities (Chapter 4).

3.2.2. Biomass Power and Fuels

Total biomass energy production has surpassed hydroelectric energy for most years since 2000 as the largest U.S. source of total renewable energy, providing 47% of renewable or 4% of total U.S. energy in 2005 (EIA, 2006). The largest source of that biomass energy (29%) was black liquor from the pulp and paper industry combusted as part of a process to recover pulping chemicals to provide process heat as well as generating electricity. Wood and wood waste from sources such as lumber mills provide more than 19% (industrial sector alone) and combusted municipal solid waste and recovered landfill gas provide about 16%, respectively, of current U.S. biomass energy (EIA, 2005d). Because energy resource generation is a byproduct of other activities in all these cases, direct impacts of climate change on these or most other sources of biomass power production derived from a waste stream may be limited unless there are significant changes in forest or agricultural productivity that are a source of the waste stream. There are few examples of literature addressing this area, though Edwards notes that climate-change-induced events such as timber die-offs could present a short-term opportunity or a long-term loss for California (Edwards, 1991).

Liquid fuel production from biomass is highly visible as a key renewable alternative to imported oil. Current U.S. production is based largely on corn for ethanol and, to a lesser extent, soybeans for biodiesel. In the longer term, cellulosic feedstocks may supplant grain and oilseed crops for transportation fuel production from biomass. Cellulosic crop residues such as corn stover and wheat straw would likely be affected by climate change the same way as the crops themselves due to a rise in average temperatures, more extreme heat days, and changes in precipitation patterns and timing, with greater impact on fuel production because that would be their primary use. Potential dedicated cellulosic energy crops for biomass fuel, such as grasses and fast-growing trees, would also be directly affected by climate change. As discussed below, limited literature suggests that for at least one region, one primary energy crop candidate—switchgrass—may benefit from climate change, both from increased temperature and increased atmospheric carbon dioxide levels.

Approximately 10% of U.S. biomass energy production (EIA, 2005d), enough to provide about 2% of U.S. transportation motor fuel (Federal Highway Administration, 2003), currently comes from ethanol made predominantly from corn grown in the Midwest (Iowa, Illinois, Nebraska, Minnesota, and South Dakota are the largest ethanol producers). Climate change sufficient to substantially affect corn production would likely impact the resource base, although production and price effects in the longer term are unclear. Production of biodiesel from soybeans—growing rapidly, but still very small—is likely a similar situation. In the long term, however, significant crop changes—and trade-offs between them as they are generally rotated with each other—would likely have an impact in the future. Looking at Missouri, Iowa, Nebraska, and Kansas, with an eye toward energy production, Brown et al., 2000 used a combination of the NCAR climate change scenario, regional climate, and crop productivity models to predict how corn, sorghum, and winter wheat (potential ethanol crops) and soybeans (biodiesel crop) would do under anticipated climate change. Negative impacts from increased temperature, positive impacts from increased precipitation, and positive impacts from increased atmospheric carbon dioxide combined to yield minimal negative change under modest carbon dioxide level increases but 5% to 12% yield increases with high carbon dioxide level increases. This assessment did not, however, account for potential impact of extreme weather events – particularly the frequency and intensity of events involving hail or prolonged droughts – that may also negatively impact energy crop production.

Although ethanol production from corn can still increase substantially (mandated to double under the recently enacted renewable fuel standard), it can still only meet a small portion of the need for renewable liquid transportation fuels to displace gasoline if dependence on petroleum imports is to be reduced. Processing the entire projected 2015 corn crop to ethanol (highly unrealistic, of course) would only yield about 35 billion gallons of ethanol, less than 14% of the gasoline energy demand projected for that year. Biomass fuel experts are counting on cellulosic biomass as the feedstock to make larger scale renewable fuel production possible. A recent joint study by the U.S. Departments of Agriculture and Energy (USDA and DOE), *Biomass as Feedstock for a Bioenergy and Bio-*

products Industry: The Technical Feasibility of a Billion-Ton Annual Supply, projected that by 2030, enough biomass could be made available to meet 40% of 2004 gasoline demand via cellulosic ethanol production and other technologies. The two largest feedstocks identified are annual crop residues and perennial dedicated energy crops (NREL, 2006).

The primary potential annual crop residues are corn stover—the leaves, stalks, and husks generally now left in the field—and wheat straw. Corn stover is the current DOE research focus in part because it is a residue with no incremental cost to grow and modest cost to harvest, but also particularly because of its potential large volume. Stover volume is roughly equivalent to grain volume, and corn is the largest U.S. agricultural crop. As such, it would be affected by climate change in much the same way as the corn crop itself, as described above.

Frequently discussed potential dedicated perennial energy crops include fast-growing trees such as hybrid poplars and willows and grasses such as switchgrass (ORNL, 2006) Switchgrass is particularly attractive because of its large regional adaptability, fast growth rate, minimal adverse environmental impact, and ease of harvesting with conventional farm equipment. The primary objective of the Brown et al., 2000 study referenced above for Missouri, Iowa, Nebraska, and Kansas was to see how climate change would affect growth of switchgrass. The study projected that switchgrass may benefit from both higher temperatures (unlike the grain crops) and higher atmospheric carbon dioxide levels, with yield increasing 74% with the modest CO_2 increase and nearly doubling with the higher CO_2 increase. Care should be taken in drawing definitive conclusions, however, from this one study. One may not expect the projected impact to be as beneficial for southern regions already warm enough for rapid switchgrass growth or more northern areas still colder than optimal even with climate change, but this analysis has not yet been conducted.

3.2.3. Wind Energy

Wind energy currently accounts for about 2.5% of U.S. renewable energy generation, but its use is growing rapidly, and it has tremendous potential due to its cost-competitiveness with fossil fuel plants for utility-scale generation and its environmental benefits. In addition, wind energy does not use or consume water to generate electricity. Unlike thermoelectric and fossil fuel generation that is inextricably linked to the availability of adequate, sustainable water supplies, wind energy can offer communities in water-stressed areas the option of economically meeting increasing energy needs without increasing demands on valuable water resources.

Although wind energy will not be impacted by changing water supplies like the other fuel sources, projected climate change impacts-- such as changes in seasonal wind patterns or strength--would likely have significant positive or negative impacts because wind energy generation is a function of the cube of the wind speed. One of the barriers slowing wind energy development today is the integration of a variable resource with the utility grid. Increased variability in wind patterns could create additional challenges for accurate wind forecasting for generation and dispatch planning and for the siting of new wind farms.

In addition to available wind resources, state and federal policy incentives have played a key role in the growth of wind energy. Texas currently produces the most wind power, followed by California, Iowa, Minnesota, Oklahoma, and Oregon (AWEA, www.awea.org/projects, 2006). These regions are expected to continue to be among the leading wind-power areas in the near term. Although North Dakota and South Dakota have modest wind development, they also have tremendous wind potential, particularly if expanded transmission capacity allows for development of sites further from major load centers.

The siting of utility-scale wind generation is highly dependent on proximity and access to the grid and the local wind speed regime. Changes in wind patterns and intensity due to climate change could have an effect on wind energy production at existing sites and planning for future development, depending on the rate and scale of that change. One study modeled wind speed change for the United States, divided into northern and southern regions under two climate-change circulation models. Overall, the Hadley Center model suggested minimal decrease in average wind speed, but the Canadian model predicted very significant decreases of 1 0%–1 5% (30%–40% decrease in power generation) by 2095. Decreases were most pronounced after 2050 in the fall for both regions and in the summer for the northern region (Breslow and Sailor, 2002).

Another study mapped wind power changes in 2050 based on the Hadley Center General Circulation Model—the one suggesting more modest change of the two used by Breslow and Sailor above. For most of the United States, this study predicted decreased wind resources by as much as 10% on an annual basis and 30% on a seasonal basis. Wind power increased for the Texas-Oklahoma region and for the Northern California-Oregon-Washington region, although the latter had decreased power in the summer. For the Northern Great Plains and for the mountainous West, however, the authors predicted decreased wind power (Segal et al. 2001). Edwards suggests that warming-induced offshore current changes could intensify summer winds for California and thus increase its wind energy potential (Edwards, 1991). Changes in diurnal wind patterns could also have a significant impact on matching of wind power production with daily load demands.

3.2.4. Solar Energy

Photovoltaic (PV) electricity generation and solar water heating are suitable for much of the United States, with current deployment primarily in off-grid locations and rooftop systems where state or local tax incentives and utility incentives are present. Utility-scale generation is most attractive in the Southwest with its high direct-radiation resource, where concentrating high-efficiency PV and solar thermal generation systems can be used. California and Arizona currently have the only existing utility-scale systems (EIA, 2005d) with additional projects being developed in Colorado, Nevada, and Arizona.

Pan et al. 2004 modeled changes to global solar radiation through the 2040s based on the Hadley Center circulation model. This study projects a solar resource reduced by as much as 20% seasonally, presumably from increased cloud cover throughout the country, but particularly in the West with its greater present resource. Increased temperature can also reduce the effectiveness of PV electrical generation and solar thermal energy collection. One international study predicts that a 2% decrease in global solar radiation will decrease solar cell output by 6% overall (Fidje and Martinsen, 2006). Anthropogenic sources of aerosols can also decrease average solar radiation, especially on a regional or localized basis. The relationship between the climate forcing effect of greenhouse gases and aerosols is complex and an area of extensive research. This field would also benefit from further analysis on the nexus between anthropogenic aerosols, climate change, solar radiation, and impacts on solar energy production.

3.2.5. Other Renewable Energy Sources

Climate change could affect geothermal energy production [6% of current U.S. renewable energy (EIA, 2005d) and concentrating solar power Rankine cycle power plants] in the same way that higher temperatures reduce the efficiency of fossil-fuel-boiler electric turbines, but there is no recent research on other potential impacts in this sector due to climate change. For a typical air-cooled binary cycle geothermal plant with a 330°F resource, power output will decrease about 1% for each 1°F rise in air temperature.

The United States currently does not make significant use of wave, tidal, or ocean thermal energy, but each of these could be affected by climate change due to changes in average water temperature, temperature gradients, salinity, sea level, wind patterns affecting wave production, and intensity and frequency of extreme weather events. Harrison observes that wave heights in the North Atlantic have been increasing and discusses how wave energy is affected by changes in wind speed (Harrison and Wallace, 2005), but very little existing research has been identified that directly addresses the potential impact of climate change on energy production from wave, tidal, or ocean thermal technologies.

3.2.6. Summary

Of the two largest U.S. renewable energy sources, hydroelectric power generation can be expected to be directly and significantly affected by climate change, while biomass power and fuel production impacts are less certain in the short term. The impact on hydroelectric production will vary by region, with potential for production decreases in key areas such as the Columbia River Basin and Northern California. Current U.S. electricity production from wind and solar energy is modest but anticipated to play a significant role in the future as the use of these technologies increases. As such, even modest impacts in key resource areas could substantially impact the cost competitiveness of these technologies due to changes in electricity production and impede the planning and financing of new wind and solar projects due to increased variability of the resource.

Renewable energy production is highly susceptible to localized and regional changes in the resource base. As a result, the greater uncertainties on regional impacts under current climate change modeling pose a significant challenge in evaluating medium to long-term impacts on renewable energy production.

3.3. Effects on Energy Transmission, Distribution, and System Infrastructure

In addition to the direct effects on operating facilities themselves, networks for transport, electric transmission, and delivery would be susceptible to changes due to climate change in stream flow, annual precipitation and seasonal patterns, storm severity, and even temperature increases (e.g., pipelines handling supercritical fluids may be impacted by greater heat loads if temperatures increase and/or cloud cover diminishes).

3.3.1. Electricity Transmission and Distribution

Severe weather events and associated flooding can cause direct disruptions in energy services. With more intense events, increased disruptions might be expected. Electricity reliability might also be affected as a result of increased demand combined with high soil temperatures and soil dryness (IPCC, 2001a). Figure 3.7 illustrates the major grid outage that was initiated by a lightning strike, as one example.

Grid technologies in use today are at least 50 years old and, although "smart grid" technologies exist, they are not often employed. Two such technologies that may be employed to help offset climate impacts include upgrading the grid by employing advanced conductors that are capable of withstanding greater temperature extremes and automation of electricity distribution (Gellings and Yeager, 2004).

3.3.2. Energy Resource Infrastructure

A substantial part of the oil imported into the United States is transported over long distances from the Middle East and Africa in supertankers. While these supertankers are able to of-fload within the ports of other countries, they are too deeply drafted to enter the shallow U.S. ports and waters. This occurs because, unlike most other countries, the continental shelf area of the eastern United States extends many miles beyond its shores and territorial waters. This leads to a number of problems related to operation of existing ports, and to programs (such as NOAA's P.O.R.T.S. Program) to improve efficiency at these ports. In addition, the Deepwater Ports Act, 1975, has led to plans to develop a number of deepwater ports either for petroleum or LNG import. These planned facilities are concentrated in relatively few locations, in particular with a concentration along the Gulf Coast (Figure 3.8). Changes in weather patterns, leading to changes in stream flows and wind speed and direction can impact operability of existing harbors. Severe weather events can impact access to deepwater facilities or might disrupt well-established navigation channels in ports where keel clearance is a concern (DOC/DOE, 2001).

Climate change may also affect the performance of the extensive pipeline system in the United States. For example, for CO_2-enhanced oil recovery, experience has shown that summer injectivity of CO_2 is about 15% less than winter injectivity into the same reservoir. The CO_2 gas temperature in Kinder Morgan pipelines during the winter is about 60°F and in late summer about 74oF. At higher temperatures, compressors and fan coolers are less efficient and are processing a warmer gas. Operators cannot pull as much gas off the supply line with the given horsepower when the CO_2 gas is warm (Source: personal communication from K. Havens of Kinder Morgan CO_2).

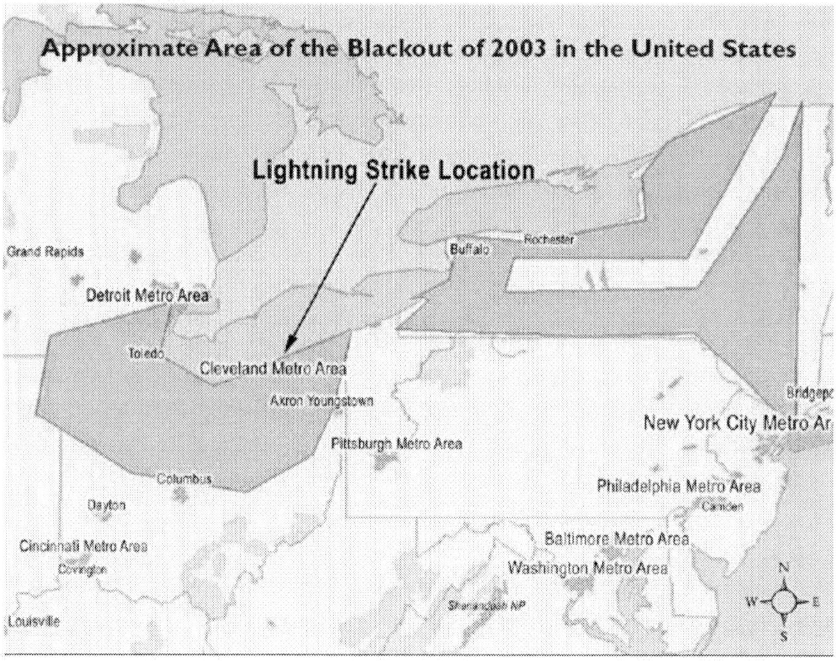

Source: NETL

Figure 3.7. Approximate Area of Blackout of 2003 in the United States.

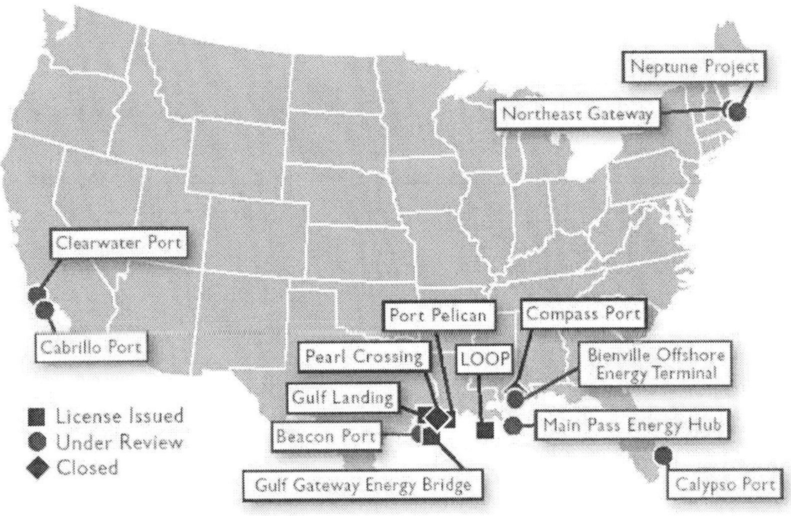

Source: U.S. Maritime Administration

Figure 3.8. Proposed Deepwater Ports For Petroleum And LNG.

Efficiencies of most gas injection are similar, and thus major gas injection projects like produced gas injection on the North Slope of Alaska have much higher gas injection and oil production during cold winter months. Persistently higher temperatures would have an impact on deliverability and injectivity for applications where the pipeline is exposed to ambient temperatures.

3.3.3. Storage and Landing Facilities

Strategic Petroleum Reserve storage locations (EIA 2004b) that are all along the Gulf Coast were selected because they provide the most flexible means for connecting to the commercial oil transport network. Figure 3.9 illustrates their locations along the Gulf Coast in areas USGS 2000 sees as being susceptible to sea- level rise, as well as severe weather events. Similarly located on the Sabine Pass is the Henry Hub, the largest gas transmission interconnection site in the U.S., connecting 14 interstate and intrastate gas transmission pipelines. Henry Hub was out of service briefly from Hurricane Katrina and for some weeks from Hurricane Rita, which made landfall at Sabine Pass.

3.3.4. Infrastructure Planning and Considerations for New Power Plant Siting

Water availability and access to coal delivery are currently critical issues in the siting of new coal-fired generation capacity. New capacity, except on coasts and large estuaries, will generally require cooling towers rather than once-through cooling water usage based on current and expected regulations (EPA, 2000) independent of climate change issues. New turbine capacity will also need to be designed to respond to the new ambient conditions.

Siting of new nuclear units will face the same water availability issues as large new coal-fired units; they will not need to deal with coal deliverability but may depend on barge transport to allow factory fabrication rather than site fabrication of large, heavy wall vessels, as well as for transportation of any wastes that need to be stored off-site.

Capacity additions and system reliability have recently become important areas for discussion. A number of approaches are being considered, such as to run auctions (or other approaches) to stimulate interest in adding new capacity, such as efforts by FERC to encourage capacity investments through regional independent system operator (ISO) organizations, without sending signals that would result in overbuilding (as has happened in the past). Planning to ensure that both predictions of needed capacity and mechanisms for stimulating companies to build such capacity (while working through the process required to announce, design, permit, and build it) will become more important as future demand is affected by climatic shifts. Similarly, site selection may need to factor in longer-term climatic changes for technologies as long-lived as coal-fired power plants (which may last for 50 - 75 years) (NARUC, 2006).

3.4. Summary of Knowledge about Possible Effects

Significant uncertainty exists about the potential impacts of climate change on energy production and distribution, in part because the timing and magnitude of climate impacts are uncertain. This report summarizes many of the key issues and provides information available on possible impacts; however this topic represents a key area for further analysis.

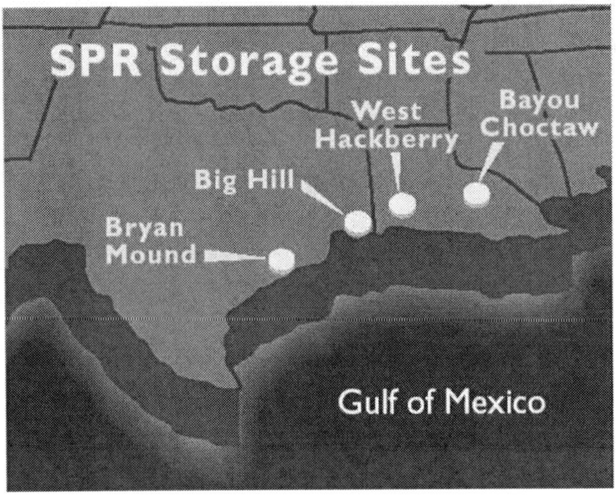

Source: NETL

Figure 3.9. Strategic Petroleum Reserve Storage Sites

Many of the technologies needed for existing energy facilities to adapt to increased temperatures and decreased water availability are available for deployment; and, although decreased efficiencies and lower output can be expected, significant disruptions seem unlikely. Incorporating potential climate impacts into the planning process for new facilities will strengthen the infrastructure. This is especially important for water resources, as electricity generation is one of many competing applications for what may be a (more) limited resource.

There are regionally important differences in adaptation needs. This is true for the spectrum of climate impacts from water availability to increased temperatures and changing patterns of severe weather events. The most salient example is for oil and gas exploration and production in Alaska, where projected temperature increases may be double the global average, and melting permafrost and changing shorelines could significantly alter the landscape and available opportunities for oil and gas production

Increased temperatures will also increase demand-side use, and the potential system-wide impacts on electricity transmission and distribution and other energy system needs are not well understood. Future planning for energy production and distribution may therefore need to accommodate possible impacts.

In: Effects of Climate Change on Energy Production... ISBN: 978-1-60741-426-1
Editor: Alrik M. Solberg

Chapter 5

POSSIBLE INDIRECT EFFECTS OF CLIMATE CHANGE ON ENERGY PRODUCTION AND USE IN THE UNITED STATES

U.S. Climate Change Science Program

INTRODUCTION

Changes in temperature, precipitation, storms, and/or sea level are likely to have direct effects on energy production and use, as summarized above; but they may also have a number of indirect effects—as climate change affects other sectors and if it shapes energy and environmental policy-making and regulatory actions (Figure 4.1). In some cases, it is possible that indirect effects could have a greater impact, positive or negative, on certain institutions and localities than direct effects.

In order to provide a basis for such a discussion, this chapter of SAP 4.5 offers a preliminary taxonomy of categories of indirect effects that may be of interest, along with a summary of existing knowledge bases about such indirect effects. Some of these effects are from climate change itself, e.g., effects on electricity prices of changing conditions for hydropower production or of more intense extreme weather events. Other effects could come from climate change related **policies** (e.g., effects of stabilization-related emission ceilings on energy prices, energy technology choices, or energy sector emissions) (Table 4.1).

Most of the existing literature is concerned with implications of climate change mitigation policies on energy technologies, prices, and emissions in the U.S. Because this literature is abundant, relatively well-known, and in some cases covered by other SAPs (such as SAP 2.2), it will be only briefly summarized here, offering links to more detailed discussions. Of greater interest to some readers may be the characterization of other possible indirect effects besides these.

4.2. Current Knowledge about Indirect Effects

4.2.1. Possible Effects on Energy Planning

Climate change is likely to affect energy planning, nationally and regionally, because it is likely to introduce new considerations and uncertainties to institutional (and individual) risk management. Such effects can arise either through anticipated changes in climate-related environmental conditions, such as hydropower potentials, possible exposure to storm damages (see Chapter 3), or changed patterns of energy demand (see Chapter 2), or through possible changes in policies and regulations.

For instance, a path-breaking study supported by EPRI and the Japanese Central Research Institute of Electric Power Industry (CRIEPI) assessed possible impacts of global climate change on six utilities, five of them in the United States (ICF, 1995). The study considered a variety of scenarios depicting a range of underlying climate, industry, and policy conditions. It found that GHG emission reduction policies could cause large increases in electricity prices, major changes in a utility's resource mix related to requirements for emission controls, and significant expansions in demand-side management programs. Major impacts are likely to be on Integrated Resource Planning regarding resource and capacity additions and/or plant retirements, along with broader implications of increased costs and prices. In another example, Burtraw et al., 2005 analyzed a nine- state northeastern regional greenhouse gas initiative (RGGI), an allowance-based regional GHG cap-and-trade program for the power sector. They found that how allowances are allocated has an effect on electricity price, consumption, and the mix of technologies used to generate electricity. Electricity prices increase in most of the cases. They also note that any policy that increases energy costs in the region is likely to cause some emission leakage to other areas outside the region as electricity generation or economic activity moves to avoid regulation and associated costs.

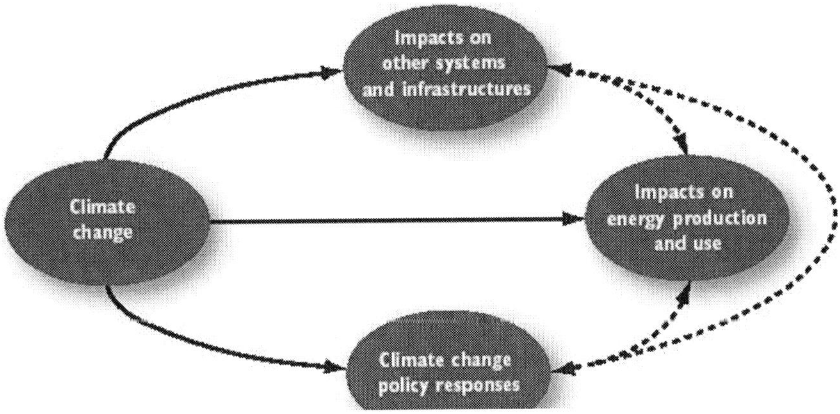

Figure 4.1. This chapter is concerned with the dashed lines in this flow diagram of connections between climate change and energy production and use

Table 4.1. Overview of the Knowledge Base about Possible Indirect Effects of Climate Change And Climate Change Policy on Energy Systems in the U.S.

Indirect Effect on Energy Systems	From Climate Change	From Climate Change Policy
On energy planning and investment	Very limited	Considerable literature
On technology R&D and preferences	Very limited	Considerable literature
On energy supply institutions	Very limited	Limited
On energy aspects of regional economies	Very limited	Some literature
On energy prices	Almost none	Considerable literature
On energy security	Almost none	Very limited
On environmental emissions from energy production/use	Very limited	Considerable literature
On energy technology/service exports	Almost none	Very limited

Electric utilities in particular are already sensitive to weather as a factor in earnings performance, and they utilize weather risk management tools to hedge against risks associated with weather-related uncertainties. Issues of interest include plans for capacity additions, system reliability assurance, and site selection for long-lived capital facilities (O'Neill, 2003). Even relatively small changes in temperature/demand can affect total capacity needs across the U.S. power sector, especially in peak periods.

Some current policy initiatives hint at what the future might be like, in terms of their possible effects on energy planning. U.S. national and state climate policy actions include a variety of traditional approaches such as funding mechanisms (incentives and disincentives); regulations (caps, codes, and standards); technical assistance (direct or in kind); research and development; information and education; and monitoring and reporting (including impact disclosure) (Rose and Zhang, 2004). Covered sectors include power generation, oil and gas, residential, commercial, industry, transportation, waste management, agriculture, and forestry. These sectors cut across private and public sector facilities and programs, as well as producers and consumers of energy (Peterson and Rose, 2006).

A variety of policy alternatives and mechanisms are described and analyzed in published literatures, including production tax credits (incorporated in the Energy Policy Act of 2005), investment tax credits, renewable energy portfolio standards, and state or regional greenhouse gas initiatives.

4.2.2. Possible Effects on Energy Production and Use *Technologies*

Perhaps the best-documented case of indirect effects of climate change on energy production and use in the United States is effects of climate change policy on technology research and development and on technology preferences and choices.

For instance, if the world moves toward concerted action to stabilize concentrations of greenhouse gases (GHG) in the earth's atmosphere, the profile of energy resources and technologies being used in the U.S. – on both the production and use sides – would have to change significantly (CCTP, 2005). Developing innovative energy technologies and approaches through science and technology research and development is widely seen as a key to reducing the role of the energy sector as a driver of climate change. Considering various climate change scenarios, researchers have modeled a number of different pathways for the world and for various regions, including the U.S., in order to inform discussions about technology options that might contribute to energy system strategies (e.g., Edmonds et al., 1996; Akimoto et al., 2004; Hoffert et al., 2002; van Vuuren et al., 2004; Kainuma et al., 2004; IPCC, 2005a; Kurosawa, 2004; Pacala and Socolow, 2004 and Paltsev et al., 2005). Recently published scenarios in CCSP SAP 2.1a, explore the U.S. implications of alternative stabilization levels of anthropogenic greenhouse gases in the atmosphere, and they explicitly consider the economic and technological foundations of such response options (CCSP, 2007a). In addition, there have been important recent developments in scenario work in the areas of non-carbon dioxide GHGs, land use and forestry emission and sinks, emissions of radiatively important non-GHGs such as black and organic carbon, and analyses of uncertainties, among many issues in increasing mitigation options and reducing costs (Nakicenovic and Riahi, 2003; IPCC, 2005b; van Vuuren et al., 2006; Weyant et al., 2006; and Placet et al., 2004).

These references indicate that an impressive amount of emissions reductions could be achieved through combinations of many different technologies, especially if diversified technology advancement is assumed. Although the full range of effects in the future is necessarily speculative, it is possible that successful development of such advanced technologies could result in potentially large economic benefits, compared with emission reductions without significant technological progress. When the costs of achieving different levels of emission reductions have been compared for cases with and without advanced technologies, many of the advanced technology scenarios projected that the cost savings from advancement would be significant (CCTP, 2005; Weyant, 2004; IPCC, 2007; CCSP, 2007a). Note, however, that there is considerable "inertia" in the nation's energy supply capital stock because institutions that have invested in expensive facilities prefer not to have them converted into "stranded assets." Note also that any kind of rapid technological transformation would be likely to have cross-commodity cost/price effects, e.g., on costs of specialized components in critical materials that are in greater demand.

4.2.3. Possible Effects on Energy Production and Use Institutions

Climate change could affect the institutional structure of energy production and use in the United States, although relatively little research has been done on such issues. Institutions include energy corporations, electric utilities, governmental organizations at all scales, and nongovernmental organizations. Their niches, size and structure, and operation tend to be sensitive to changes in "market" conditions from any of a variety of driving forces, these days including such forces as globalization, technological change, and social/cultural change (e.g., changes in consumer preferences). Climate change is likely to interact with other driving forces in ways that could affect institutions concerned with energy production and use.

Most of the very limited research attention to this type of effect has been focused on effects of climate change policy (e.g., policy actions to reduce greenhouse gas emissions) on U.S. energy institutions, such as on the financial viability of U.S. electric utilities (see, for instance, WWF, 2003). Other effects could emerge from changes in energy resource/technology mixes due to climate change: e.g., changes in renewable energy resources and costs or changes in energy R&D investment patterns.

Most of these issues are speculative at this time, but identifying them is useful as a basis for further discussion. Issues would appear to include effects on planning, above.

4.2.3.1. Effects on the Institutional Structure of the Energy Industry

Depending on its impacts, climate change could encourage large energy firms to move into renewable energy areas that have been largely the province of smaller firms, as was the case in some instances in the wake of the energy "shocks" of the 1970s (e.g., Flavin and Lenssen, 1994). This kind of diversification into other "clean energy" fields could be reflected in horizontal and/or vertical integration. Possible effects of climate change on these and other institutional issues (such as organizational consolidation vs fragmentation) have not been addressed systematically in the research literature; but some large energy firms are exploring a wider range of energy technologies and some large multinational energy technology providers are diversifying their product lines to be prepared for possible changes in market conditions.

4.2.3.2. Effects on Electric Utility Restructuring

Recent trends in electric utility restructuring have included increasing competition in an open electricity supply marketplace, which has sharpened attention to keeping O&M costs for infrastructure as low as possible. Some research literature suggests that one side-effect of restructuring has been a reduced willingness on the part of some utilities to invest in environmental protection beyond what is absolutely required by law and regulation (Parker, 1999; Senate of Texas, 1999), although this issue needs further study. If climate change introduces new risks for utility investment planning and reliability, it is possible that policies and practices could encourage greater cooperation and collaboration among utilities.

4.2.3.3. Effects on the Health of Fossil Fuel-Related Industries

If climate change is associated with policy and associated market signals that decarbonization of energy systems, industries focused on the production of fossil fuels, converting them into useful energy forms, transporting them to demand centers, and providing them to users could face shrinking markets and profits. The coal industry seems especially endangered in such an eventuality. In the longer run, this type of effect depends considerably on technological change: e.g., affordable carbon capture and sequestration, fuel cells, and efficiency improvement. It is possible that industries (and regions) concentrated on fossil fuel extraction, processing, and use will seek to diversify as a hedge against risks of economic threats from climate change policy.

4.2.3.4. Effects on Other Supporting Institutions Such as Financial and Insurance Industries

Many major financial and insurance institutions are gearing up to underwrite emission trading contracts, derivatives and hedging products, wind and biofuel crop guarantee covers for renewable energy, and other new financial products to support carbon emission trading, while they are concerned about exposure to financial risks associated with climate change impacts. In recent years, various organizations have tried to engage the global insurance industry in the climate change debate. Casualty insurers are concerned about possible litigation against companies responsible for excessive GHG emissions, and property insurers are concerned about future uncertainties in weather damage losses. However, it is in the field of adaptation where insurers are most active, and have most to contribute. Two hundred major companies in the financial sector around the world have signed up to the UN Environment Program's - Finance Initiative, and 95 institutional investment companies have so far signed up to the Carbon Disclosure Project. They ask businesses to disclose investment-relevant information concerning their GHGs. Their website provides a comprehensive registry of GHGs from public corporations. More than 300 of the 500 largest companies in the world now report their emissions on this website, recognizing that institutional investors regard this information as important for shareholders (Crichton, 2005).

4.3. Possible Effects on Energy-Related Dimensions of Regional and National Economies

It is at least possible that climate change could have an effect on regional economies by impacting regional comparative advantages related to energy availability and cost. Examples could include regional economies closely associated with fossil fuel production and use (especially coal) if climate change policies encourage decarbonization, regional economies dependent on affordable electricity from hydropower if water supplies decrease or increase, regional economies closely tied to coastal energy facilities that could be threatened by more intense coastal storms (Chapter 3), and regional economies dependent on abundant electricity supplies if demands on current capacities increase or decrease due to climate change.

Attempts to estimate the economic impacts that could occur 50–100 years in the future have been made using various climate scenarios, but the interaction of climate and the nation's economy remains very difficult to define. Most studies of the economic impacts of global warming have analyzed the impacts on specific sectors (such as agriculture) or on regional ecosystems (e.g. Fankhauser, 1995; Mendelsohn and Neumann, 1999; Nordhaus and Boyer, 2000; Mendelsohn et al., 1994; Tol, 2002; Nordhaus, 2006). However, not many impact studies have concentrated on the energy sector. Significant uncertainties therefore surround projections of climate change induced energy sector impacts on the U.S. or regional economies. Changnon estimated that annual national economic losses from the energy sector will outweigh the gains in years with major weather and climate extremes (Changnon, 2005). Jorgenson et al., 2004, studied impacts of climate change on various sectors of the U.S. economy from 2000 – 2 100. In three optimistic scenarios, they conclude that increased energy availability and cost savings from reduced natural gas-based space heating more than compensate for increased expenditures on electricity-based space cooling. These unit cost reductions appear as productivity increases and, thus, improve the economy, whereas other three pessimistic scenarios show that electricity-based space conditioning experiences relatively larger productivity losses than does space conditioning from coal, wood, petroleum or natural gas; accordingly its (direct) unit cost rises faster and thus produces no benefits to the economy. Additionally, higher domestic prices discourage exports and promote imports leading to a worsening real trade balance. According to Mendelsohn et al., 2000, the U.S. economy could benefit from the climate change induced energy sector changes. However, Mendelsohn and Williams, 2004 suggest that climate change will cause economic damages in the energy sector in every scenario. They suggest that temperature changes cause most of the energy impacts. Larger temperature increases generate significantly larger economic damages. The damages are from increased cooling expenditures required to maintain desired indoor temperatures. In the empirical studies, these cost increases outweighed benefits of the reduced heating expenditures unless starting climates are very cool (Mendelsohn and Neumann, 1999; Mendelsohn, 2001) (also see Chapter 2).

In California, a preliminary assessment of the macroeconomic impacts associated with the climate change emission reduction strategies (CEPA, 2006) shows that, while some impacts on the economy could be positive if strategies reduce energy costs, other impacts might be less positive. For example, the study emphasizes that even relatively small changes in in-state hydropower generation result in substantial extra expenditure burdens on an economy for energy generation, because losses in this "free" generation must be purchased from other sources; for example, a 10% decrease in hydroelectric supply would impose a cost of approximately $350 million in additional electricity expenditures annually (Franco and Sanstad, 2006). Whereas electricity demand is projected to rise in California between 3 to 20 % by the end of this century, peak electricity demand would increase at a faster rate. Since annual expenditures of electricity demand in California represent about $28 billion, even such a relatively small increase in energy demand would result in substantial extra energy expenditures for energy services in the state; a 3 % increase in electricity demand by 2020 would translate into about $930 million (in 2000 dollars) in additional electricity expenditures (Franco and Sanstad, 2006). Particular concerns are likely to exist in areas where summer electricity loads already strain supply capacities (e.g., Hill and Goldberg 2001; Kelly et al. 2005; Rosenzweig and Solecki, 2001) and where transmission and distribution networks have limited capacities to adapt to changes in regional demands, especially seasonally (e.g., London Climate Change, Partnership 2002).

Rose and others have examined effects of a number of climate change mitigation policies on U.S. regions in general and the Susquehanna River basin in particular (Rose and Oladosu, 2002; Rose and Zhang, 2004; Rose et al., 1999; Rose et al., 2006). In general, they find that such policy options as emission permits tradable among U.S. regions might have less than expected effects, with burdens impacting at least one Southern region that needs maximum permits but whose economy is not among the nation's strongest. Additionally, they discuss Pennsylvania's heavy reliance on coal production and use infrastructure that increases the price of internal carbon dioxide mitigation. They suggest that the anomalies stem from the fact that new entrants, like Pennsylvania, into regional coalitions for cap-and-trade configuration may raise the permit price, may undercut existing states' permit sales, and may be able to exercise market power. Particularly, they raise an issue of the "responsibility" for emissions. Should fossil fuel producing regions take the full blame for emissions, or are the using regions also responsible? They find that aggregate impacts of a carbon tax on the Susquehanna River Basin would be negative but quite modest.

Concerns remain, however, that aggressive climate policy interventions to reduce GHG emissions could negatively affect regional economies linked to coal and other fossil energy production. Concerns also exist that climate change itself could affect the economies of areas exposed to severe weather events (positively or negatively) and areas whose economies are closely linked to hydropower and other aspects of the "energy-water nexus."

4.4. Possible Relationships with Other Energy-Related Issues

Many other types of indirect effects are possible, although relatively few have received research attention. Without asserting that this listing is comprehensive, such effects might include the following types.

4.4.1. Effects of Climate Change in Other Countries on U.S. Energy Production and Use

We know from recent experience that climate variability outside the U.S. can affect energy conditions in the U.S.; an example is an unusually dry year in Spain in 2005 that led the country to enter the international LNG market to compensate for scarce hydropower, which in turn raised LNG prices for U.S. consumption (Alexander's Gas & Oil Connections, 2005). It is important, therefore, to consider possible effects of climate change not only on international energy product suppliers and international energy technology buyers but also on other countries whose participation in international markets could affect U.S. energy availability and prices from international sources, which could have implications for energy security (see below). Climate change-related energy supply and price effects could be coupled with other price effects of international trends on U.S. energy, infrastructures, such as effects of aggressive programs of infrastructure development on China and India.

As indicated in Chapter 2, a particularly important case is U.S. energy inputs from Canada. Canada is the largest single source of petroleum imports by the U.S. (about 2.2 million barrels per day) and exports more than 15% of the natural gas consumed in the U.S. (EIA 2005a, 2006). In 2004, it exported to the U.S. 33 MWh of electricity, compared with imports of 22.5 MWh (EIA, 2005b). Climate change could affect electricity exports and imports, for instance if electricity demands for space cooling increase in Canada or if climate change affects hydropower production in that country.

4.4.2. Effects of Climate Change on Energy Prices*

A principal mechanism in reducing vulnerabilities to climate-related (and other) changes potentially affecting the energy sector is the operation of the energy market, where price variation is a key driver. Effects of climate change on energy prices are in fact interwoven with effects of energy prices on risk management strategies, in a dynamic that could work in both directions at once; and it would be useful to know more about roles of energy markets in reducing vulnerabilities to climate change impacts, along with possible adaptations in the functioning of those markets. Although price effects of climate change itself are not analyzed in the literature, aside from effects of extreme events such as Hurricane Katrina, substantial research has been done on possible energy price effects of greenhouse gas emission reductions.

Estimates of costs of emission reduction vary widely according to assumptions about such issues as how welfare is measured, ancillary benefits, and effects in stimulating technological innovation; and therefore any particular set of cost estimates includes considerable uncertainty. According to an Interlaboratory Working Group (IWG, 2000), benefits of emission reduction would be comparable to costs, and the National Commission on Energy Policy 2004 estimates that its recommended policy initiatives would be, on the whole, revenue-neutral with respect to the federal budget. Other participants in energy policymaking, however, are convinced that truly significant carbon emission reductions would have substantial economic impacts (GAO, 2004).

Globally, IPCC, 2001 projected that total CO_2 emissions from energy supply and conversion could be reduced in 2020 by 350 to 700 Mt C equivalents per year, based on options that could be adopted through the use of generally accepted policies, generally at a positive direct cost of less than U.S.$100 per t C equivalents. Based on DOE/EIA analyses in 2000, this study includes estimates of the cost of a range of specific emission-reducing technologies for power generation, compared with coal-fired power, although the degree of uncertainty is not clear. Within the United States, the report estimated that the cost of emission reduction per metric ton of carbon emissions reduced would range from –$170 to +$880, depending on the technology used. Marginal abatement costs for the total United States economy, in 1990 U.S. dollars per metric ton carbon, were estimated by a variety of models compared by the Energy Modeling Forum at $76 to $410 with no emission trading, $14 to $224 with Annex I trading, and $5 to $123 with global trading.

Similarly, the National Commission on Energy Policy 2004 considered costs associated with a tradable emission permit system that would reduce United States national greenhouse gas emission growth from 44% to 33% from 2002 to 2025, a reduction of 760 Mt CO_2 (207 Mt C) in 2025 compared with a reference case. The cost would be a roughly 5% increase in total end-use expenditures compared with the reference case. Electricity prices would rise by 5.4% for residential users, 6.2% for commercial users, and 7.6% for industrial users.

The IWG 2000 estimated that a domestic carbon trading system with a $25/t C permit price would reduce emissions by 13% compared with a reference case, or 230 Mt CO_2 (63 Mt C), while a $50 price would reduce emissions by 17 to 19%, or 306 to 332 Mt CO_2 (83-91 Mt C). Both cases assume a doubling of United States government appropriations for cost-shared clean energy research, design, and development.

Net costs to the consumer, however, are balanced in some analyses by benefits from advanced technologies that are developed and deployed on an accelerated schedule due to policy interventions and changing public preferences. The U.S. Climate Change Technology Program, 2005: pp. 3–19, illustrates how costs of achieving different stabilization levels can conceivably be reduced substantially by the use of advanced technologies, and IWG (2000) estimates that net end-user costs of energy can actually be reduced by a domestic carbon trading system if it accelerates the market penetration of more energy-efficient technologies (see Section 4.2.2 above).

4.4.3. Effects of Climate Change on Environmental Emissions

Climate change is very likely to lead to reductions in environmental emissions from energy production and use in the U.S., although possible effects of climate change responses are complex. For instance, cap and trade policy responses might not translate directly into lower total emissions. In general, however, the available research literature indicates that climate change policy will affect choices of energy resources and technologies in ways that, overall, reduce greenhouse gas and other environmental emissions (see indirect impacts on technologies above).

4.4.4. Effects of Climate Change on Energy Security

Climate change relates to energy security because different drivers of energy policy interact. As one example, some strategies to reduce oil import dependence, such as increased use of renewable energy sources in the U.S., are similar to strategies to reduce GHG emissions as a climate change response (e.g., IEA, 2004; O'Keefe, 2005). Other strategies such as increased domestic fossil fuel production and use could be contradictory to climate change policies. The complexity of connections between climate change responses and energy security concerns can be illustrated by choices between uses of biomass to reduce fossil fuel use in electricity generation, a priority for net greenhouse gas emissions, and uses of biomass to displace oil and gas imports, a priority for energy security policy. Although the relative effects of the two options are not entirely unrelated (i.e., both could have some effect in reducing oil and gas imports and both could have some effect in reducing net greenhouse gas emissions), the balance in contributions to these two policy priorities would be different.

As another example, energy security relates not only to import dependence but also to energy system reliability, which can be threatened by possible increases in the intensity of severe weather events. A different kind of issue is potential impacts of abrupt climate change in the longer run. One study has suggested that abrupt climate change could lead to very serious international security threats, including threats of global energy crises, as countries act to defend and secure supplies of essential commodities (Schwarz and Randall, 2004). Clearly, then, relationships between climate change response and energy security are complex, but they are potentially important enough to deserve further study.

4.4.5. Effects of Climate Change on Energy Technology and Service Exports

Finally, climate change could affect U.S. energy technology and service exports. It is very likely that climate change will have some impacts on global energy technology, institutional, and policy choices. Effects of these changes on U.S. exports would probably be determined by whether the U.S. is a leader or a follower in energy technology and policy responses to concerns about climate change. More broadly, carbon emission abatement actions by various countries are likely to affect international energy flows and trade flows in energy technology and services (e.g., Rutherford, 2001). In particular, one might expect flows of carbon-intensive energy forms and energy technologies and energy-intensive products to be affected.

4.5. SUMMARY OF KNOWLEDGE ABOUT INDIRECT EFFECTS

Regarding indirect effects of climate change on energy production and use in the United States, the available research literature tells us the most about possible changes in energy resource/technology preferences and investments, along with associated reductions in GHG emissions and effects on energy prices. Less-studied but also potentially important are possible impacts on the institutional structure of energy supply in the United States, responding to changes in perceived investment risks and emerging market and policy realities, and possible interactions between energy prices and roles of energy markets in managing risks and reducing vulnerabilities. Perhaps the most important insight from the limited current research literature is that climate change will affect energy production and use not only as a driving force in its own right but in its interactions with other driving forces such as energy security. Where climate change response strategies correspond with other issue response strategies, they can add force to actions such as increased reliance on domestic noncarbon energy supply sources. Where climate change impacts contradict other driving forces for energy decisions, it is much less clear what their net effect would be on energy production and use.

End Notes

* Adapted in part from *CCSP SAP 2.2, State of the Carbon Cycle Report,* Chapter 6, "Energy Conversion."

In: Effects of Climate Change on Energy Production… ISBN: 978-1-60741-426-1
Editor: Alrik M. Solberg

Chapter 6

Conclusions and Research Priorities

U.S. Climate Change Science Program

Introduction

The previous chapters have summarized a variety of currently available information about effects of climate change on energy production and use in the United States. For two reasons, it is important to be careful about drawing firm conclusions about effects at this time. One reason is that the research literatures on many of the key issues are limited, supporting an identification of issues but not a resolution of most uncertainties. A second reason is that, as with many other categories of climate change effects in the U.S., the effects depend on a wide range of factors beyond climate change alone, such as patterns of economic growth and land use, patterns of population growth and distribution, technological change, and social and cultural trends that could shape policies and actions, individually and institutionally.

Accordingly, this final chapter of SAP 4.5 will sketch out what appear, based on the current knowledge base, to be the most likely types of effects on the energy sector. These should be considered along with effects on other sectors that should be considered in risk management discussions in the near term. As indicated in Chapter 1, conclusions are related to degrees of likelihood: likely (2 chances out of 3), very likely (9 chances out of 10), or virtually certain (99 chances out of 100). The chapter will then discuss issues related to prospects for energy systems in the U.S. to adapt to such effects, although literatures on adaptation are very limited. Finally, it will suggest a limited number of needs for expanding the knowledge base so that, when further assessments on this topic are carried out, conclusions about effects can be offered with a higher level of confidence.

CONCLUSIONS ABOUT EFFECTS

Based on currently available projections of climate change in the United States, a number of conclusions can be suggested about likely effects on energy *use* in the U.S. over a period of time addressed by the research literature (near to midterm). Long-term conclusions are difficult due to uncertainties about such driving forces as technological, change, institutional change, and climate change policy responses.

- Climate change will mean reductions in total U.S. energy demand for space heating for buildings, with effects differing by region (virtually certain).
- Climate change will mean increases in total U.S. energy demand for space cooling, with effects differing by region (virtually certain).
- Net effects on energy use will differ by region, with net lower total energy requirements for buildings in net heating load areas and net higher energy requirements in net cooling load areas, with overall impacts affected by patterns of interregional migration – which are likely to be in the direction of net cooling load regions – and investments in new building stock (virtually certain).
- Temperature increases will be associated with increased peak demands for electricity (very likely).
- Other effects of climate change are less clear, but some could be nontrivial: e.g., increased energy use for water pumping and/or desalination in areas that see reductions in water supply (very likely).
- Lower winter energy demands in Canada could add to available electricity supplies for a few U.S. regions (likely).

A number of conclusions can be offered with relatively high levels of confidence about effects of climate change on energy *production and supply* in the U.S., but generally the research evidence is not as strong as for effects on energy use:

- Changes in the distribution of water availability in the U.S. will affect power plants; in areas with decreased water availability, competition for water supplies will increase between energy production and other sectors (virtually certain).
- Temperature increases will decrease overall thermoelectric power generation efficiency (virtually certain).
- In some regions, energy resource production and delivery systems are vulnerable to effects of sea level rise and extreme weather events, especially the Gulf Coast and the East Coast (virtually certain).
- In some areas, the siting of new energy facilities and systems could face increased restrictions, related partly to complex interactions among the wider range of water uses as well as sea-level rise and extreme event exposures (likely
- Incorporating possible climate change impacts into planning processes could strengthen energy production and distribution system infrastructures, especially regarding water resource management (likely).
- Hydropower production is expected to be directly and significantly affected by climate change, especially in the West and Northwest (very likely).
- Climate change is expected to mean greater variability in wind resources and direct solar radiation, substantially impacting the planning, siting, and financing of these technologies (likely).
- Increased temperatures and other climate change effects will affect energy transmission and distribution requirements, but these effects are not well-understood.

Overall, the current energy supply infrastructure is often located in areas where climate change impacts might occur, but large-scale disruptions are not likely except during extreme weather events. Most of the effects on fossil and nuclear electricity components are likely to be modest changes in water availability and/or cycle efficiency.

California is one U.S. state where impacts on both energy use and energy production have been studied with some care (See Box 5.1: California: A Case Study).

About *indirect effects of climate change on energy production and use* in the U.S., conclusions are notably mixed. Conclusions related to possible impacts of climate change policy interventions on technology choice and emissions can be offered with relatively high confidence based on published research:

- Climate change concerns are very likely to affect perceptions and practices related to risk management behavior in investment by energy institutions (very likely).
- Climate change concerns, especially if they are expressed through policy interventions, are almost certain to affect public and private sector energy technology R&D investments and energy resource/technology choices by energy institutions, along with associated emissions (virtually certain).
- Climate change can be expected to affect other countries in ways that in turn affect U.S. energy conditions (very likely).

BOX 5.1 CALIFORNIA: A CASE STUDY

California is unique in the United States as a state that has examined possible effects of climate change on its energy production and use in some detail (also see Box 2.2). Led by the California Energy Commission and supported by such nearby partners as the Electric Power Research Institute, the University of California–Berkeley, and the Scripps Institution of Oceanography, the state is developing a knowledge base on this subject that could be a model for other states and regions (as well as the nation as a whole).

Generally, the analyses to date (many of which are referenced in Chapters 2 and 3) indicate that electricity demand will grow due to climate change, with an especially close relationship between peak electricity demand and temperature increases (Franco and Sanstad 2006), and water supply – as an element of the "energy-water nexus" – will be affected by a reduction in the Sierra snowpack (by as much as 70-90 % over the coming century: Vicuña et al. 2006). Patterns of urbanization could add to pressures for further energy supplies. Adaptations to these and other climate change impacts appear possible, but they could be costly (Franco 2005). Overall economic impacts will depend considerably on the effectiveness of response measures, which tend currently to emphasize emission reduction but also consider impact scenarios and potential adaptation measures (CEPA 2006).

Other relevant studies of the California context for climate change effects reinforce an impression that effects of warming and snowpack reduction could be serious (Hayhoe et al. 2004) and that other ecosystems related to renewable energy potentials could be affected as well (Union of Concerned Scientists 1999).

Other types of possible indirect effects can be suggested as a basis for discussion, but conclusions must await further research:

- Climate change effects on energy production and use could in turn affect some regional economies, either positively or negatively (likely).
- Climate change may have some effects on energy prices in the U.S., especially associated with extreme weather events (very likely).
- Climate change concerns are likely to interact with some driving forces behind policies focused on U.S. energy security, such as reduced reliance on conventional petroleum products (likely).

These conclusions add up to a picture that is cautionary rather than alarming. Since in many cases effects that could be a concern to U.S. citizens and U.S. energy institutions are some decades in the future, there is time to consider strategies for adaptation to reduce possible negative impacts and take advantage of possible positive impacts.

5.3. Considering Prospects for Adaptation

The existing research literature tends to treat the U.S. energy sector mainly as a *driving force* for climate change rather than a sector *subject to impacts* from climate change. As a result, there is very little literature on adaptation of the energy sector to effects of climate change, and the following discussion is therefore largely speculative.

Generally, both energy users and providers in the U.S. are accustomed to changes in conditions that affect their decisions. Users see energy prices fluctuate with international oil market conditions and with Gulf Coast storm behavior, and they see energy availability subject to short-term shortages for a variety of reasons (e.g., the California energy market crises of 2000/200 1 or electricity blackouts in some Northeastern cities in 2003). Energy providers cope with shifting global market conditions, policy changes, financial variables such as interest rates for capital infrastructure lending, and climate variability. In many ways, the energy sector is among the most resilient of all U.S. economic sectors, at least in terms of responding to changes within the range of historical experience. For instance, electric utilities routinely consider planning and investment strategies that consider weather variables (Niemeyer, 2005); and one important guide to adaptation to climate *change* is what makes sense in adapting to climate *variability* (Franco 2005).

On the other hand, such recent events as Hurricane Katrina (Box 5.2: Hurricane Katrina and the Gulf Coast: A Case Study) suggest that the U.S. energy sector is better at responding to relatively short-term variations and uncertainties than to changes that reach beyond the range of familiar short-term variabilities (Niemeyer 2005). In fact, the confidence of U.S. energy institutions about their ability to reduce exposure to risks from short-term variations might tend to reduce their resilience to larger long-term changes, unless an awareness of risks from such long-term changes is heightened.

Box 5.2 Hurricane Katrina and the Gulf Coast: A Case Study

It is not possible to attribute the occurrence of Hurricane Katrina, August 29, 2005, to climate change; but projections of climate change say that extreme weather events are very likely to become more intense. If so (e.g., more of the annual hurricanes at higher levels of wind speed and potential damages), then the impacts of Katrina are an indicator of possible impacts of one manifestation of climate change.

Impacts of Katrina on energy systems in the region and the nation were dramatic at the time, and some impacts remained many months later. The hurricane itself impacted coastal and offshore oil and gas production, offshore oil port operation (stopping imports of more than one million bbl/d of crude oil), and crude oil refining along the Louisiana Gulf Coast (Figures 3.4 a-d). Within only a few days, oil product and natural gas prices had risen significantly across the U.S. As of mid-December 2005, substantial oil and gas production was still shut-in, and refinery shutdowns still totalled 367, 000 bbl/d (EIA 2005) (see Chapter 3).

Possibilities for adaptation to reduce risks of damages from future Katrinas are unclear. They might include such alternatives as hardening offshore platforms and coastal facilities to be more resilient to high winds, wave action, and flooding (potentially expensive) and shifting the locations of some coastal refining and distribution facilities to less vulnerable sites, reducing their concentration in the Gulf Coast. (potentially very expensive).

Adaptations to effects of climate change on energy *use* may focus on increased demands for space cooling in areas affected by warming and associated increases in total energy consumption costs. Alternatives could include reducing costs of cooling for users through energy efficiency improvement in cooling equipment and building envelopes; responding to likely increases in demands for electricity for cooling through expanded generation capacities, expanded interties, and possibly increased capac ities for storage; and responding to concerns about increased peak demand in electricity loads, especially seasonally, through contingency planning for load-leveling. Over a period of several decades, for instance, technologies are likely to respond to consumer concerns about higher energy bills where they occur.

Many technologies that can enable adaptations to effects on energy *production and supply* are available for deployment. The most likely adaptation in the near term is an increase in perceptions of uncertainty and risk in longer-term strategic planning and investment, which could seek to reduce risks through such approaches as diversifying supply sources and technologies and risk-sharing arrangements.

Adaptation to *indirect effects* of climate change on the energy sector is likely to be bundled with adaptation to other issues for energy policy and decision-making in the U.S., such as energy security: for instance, in the development of lower carbon-emitting fossil fuel use technology ensembles, increased deployment of renewable energy technologies, and the development of alternatives to fossil fuels and effects on energy institutional structures. Issues related to effects of climate change on other countries linked with U.S. energy conditions are likely to be addressed through attention by both the public and private sectors to related information systems and market signals.

It seems possible that adaptation challenges would be greatest in connection with possible increases in the intensity of extreme weather events and possible significant changes in regional water supply regimes. More generally, adaptation prospects appear to be related to the magnitude and rate of climate change (e.g., how much the average temperature rises before stabilization is achieved, how rapidly it moves to that level, and how variable the climate is at that level), with adaptation more likely to be able to cope with effects of lesser amounts, slower rates of change, and less variable climate (Wilbanks et al., 2007).

Generally, prospects for these types of adaptations depend considerably on the level of awareness of possible climate changes at a relatively localized scale and possible implications for energy production and use – the topic of this study. When the current knowledge base to support such awareness is so limited, this suggests that expanding the knowledge base is important to the energy sector in the United States.

5.4. Needs for Expanding the Knowledge Base

Expanding the knowledge base about effects of climate change on energy production and use in the United States is not just a responsibility of the federal government. As the work of such institutions as the Electric Power Research Institute and the California Energy Commission demonstrates, a wide variety of parts of U.S. society have knowledge, expertise, and data to contribute to what should be a broad-based multi-institutional collaboration.

Recognizing that roles in these regards will differ among federal and state governments, industry, nongovernmental institutions, and academia and that all parties should be involved in discussions about how to proceed, this study suggests the following needs for expanding the knowledge base on its topic, some of which are rooted in broader needs for advances in climate change science.

5.4.1 General Needs

- Improved capacities to project climate change and its effects on a relatively fine-grained geographic scale, especially of precipitation changes and severe weather events: e.g., in order to support evaluations of impacts at local and small-regional scales, not only in terms of gradual changes but also in terms of extremes, since many energy facility decisions are made at a relatively localized scale;
- Research on and assessments of implications of extreme weather events for energy system resiliency, including strategies for both reducing and recovering from impacts;
- Research on and assessments of potentials, costs, and limits of adaptation to risks of adverse effects, for both supply and use infrastructures;
- Research on efficiency of energy use in the context of climate warming, with an emphasis on technologies and practices that save cooling energy and reduce electrical peak load;
- Research on and assessments of implications of changing regional patterns of energy use for regional energy supply institutions and consumers;
- Improvements in the understanding of effects of changing conditions for renewable energy and fossil energy development and market penetration on regional energy balances and their relationships with regional economies;

- In particular, improvements in understanding likely effects of climate change in Arctic regions and on storm intensity to guide applications of existing technologies and the development and deployment of new technologies and other adaptations for energy infrastructure and energy exploration and production in these relatively vulnerable regions; and
- Attention to linkages and feedbacks among climate change effects, adaptation, and mitigation; to linkages between effects at different geographic scales; and relationships between possible energy effects and other possible economic, environmental, and institutional changes (Parson et al., 2003; Wilbanks, 2005).

5.4.2. Needs Related to Major Technology Areas

- Improving the understanding of potentials to increase efficiency improvements in space cooling;
- Improving information about interactions among water demands and uses where the quantity and timing of surface water discharge is affected by climate change;
- Improving the understanding of potential climate change and localized variability on energy production from wind and solar technologies;
- Developing strategies to increase the resilience of coastal and offshore oil and gas production and distribution systems to extreme weather events;
- Pursuing strategies and improved technology potentials for adding resilience to energy supply systems that may be subject to stress under possible scenarios for climate change;
- Improving understandings of potentials to improve resilience in electricity supply systems through regional intertie capacities and distributed generation; and
- Research on and assessments of the impacts of severe weather events on sub-sea pipeline systems, especially in the Gulf of Mexico, and strategies for reducing such impacts.[1]

Other needs for research exist as well, and the process of learning more about this topic in coming years may change perceptions of needs and priorities; but based on current knowledge, these appear to be high priorities in the next several years.

End Notes

[1] Note that CCSP SAP 4.7, The Impacts of Climate Change on Transportation: A Gulf Coast Study, considers imacts on pipelines and other transportation infrastructures in the Gulf Coast region (CCSP, 2007b).

In: Effects of Climate Change on Energy Production... ISBN: 978-1-60741-426-1
Editor: Alrik M. Solberg

Chapter 7

REFERENCES

U.S. Climate Change Science Program

ACIA (2004). *Impacts of a Warming Arctic – Arctic Climate Impact Assessment*, Cambridge: Cambridge University Press.

Akimoto, K., Tomoda, T., Fujii, Y. & Yamaji, K. (2004). Assessment of global warming mitigation options with integrated assessment model DNE21, *Energy Economics 26*, 635– 653.

Alaska Department of Natural Resources (2004). *Tundra Travel Modeling Project*, p. 2.

Alexander's Gas & Oil Connections (2005). European gas profiles, *Market Reports*, November 2005.

Amato, A. D., Ruth, M., Kirshen P. & Horwitz, J. (2005). Regional energy demand responses to climate change: Methodology and application to the Commonwealth of Massachusetts. *Climatic Change*, *71*, 175–201.

API (2006). *Recommended Practice,* 95F, First Edition, Washington: American Petroleum Institute.

API (2006a). *Recommended Practice*, 95J, First Edition, Washington: American Petroleum Institute.

API (2006b). *Gulf of Mexico Jackup Operations for Hurricane Season—Interim Recommendations*, Washington: American Petroleum Institute

API (2006c). *Interim Guidance for Gulf of Mexico MODU Mooring Practice—2006 Hurricane Season*. Washington: American Petroleum Institute.

Aspen Environmental Group and M-Cubed (2005). *Potential Changes In Hydropower Production From Global Climate Change In California And the Western United States*, CEC700-2005-0 10. California Energy Commission, Sacramento, California.

Atkinson, B. A., Barnaby, C. S., Wexler, A. H. & Wilcox, B. A. (1981). *Proceedings of the Annual Meeting – American Section of the International Solar Energy Society, Vol./Issue: 6*; National Passive Solar Conference; September 8, 1981; Portland, OR.

Badri, M. A. (1992). *Analysis of demand for electricity in the United States, Energy 17(7)*, 725–73 3.

Barnett, T. et al. (2004). The effects of climate change on water resources in the west: Introduction and overview, *Climatic Change, 62*, 1-11.

Baxter, L. W. & Calandri, K. (1992). Global warming and electricity demand: A study of California. *Energy Policy*, *20(3)*, 233–244.

Belzer, D. B., Scott, M. J. & Sands, R. D. (1996). Climate change impacts on U.S. commercial building energy consumption: an analysis using sample survey data, *Energy Sources 18(2)*, 177–201.

Billings Gazette (2005). *South Dakota Governor Seeks Summit on Missouri River*, February 2005. Accessed at: http://www:Billingsgazette.com

Breslow, P. & Sailor, D. (2002). Vulnerability of wind power resources to climate change in the continental United States, *Renewable Energy*, *27*,. 585–598.

Brown, R. A. et al. (2000). Potential production and environmental effects of switchgrass and traditional crops under current and greenhouse-altered climate in the central United States: A simulation study, Agriculture, *Ecosystems, and Environment*, *78*, 31-47.

Burtraw, B., Palmer, K. & Kahn, D. (2005). *Allocation of CO2 Emissions Allowances in the Regional Greenhouse Gas Capand-Trade Program*, RFF DP 05-25, Washington, DC: Resources for the Future.

California Energy Commission (2006). Historic State-Wide California Electricity Demand. Accessed at http://energy

CCSP (2007a). *Scenarios of Greenhouse Gas Emissions and Atmospheric Concentrations*. Sub-report 2.1A of Synthesis and Assessment Product 2.1 by the U.S. Climate Change Science Program and the Subcommittee on Global Change Research. Department of Energy, Washington, DC, 154 pp.

CCSP (2007b). *The Impacts of Climate Change on Transportation: A Gulf Coast Study*. Synthesis and Assessment Product 4.7 by the U.S. Climate Change Science Program and the Subcommittee on Global Change Research. Department of Transportation, Washington, DC, forthcoming.

CCTP (2005). *U.S. Climate Change Technology Program: Vision and Framework for Strategy and Planning*, U.S. Climate Change Technology Program, Washington, U.S. Climate Change Technology Program.

CEPA (2006). *Report to the Governor and Legislature. Climate Action Team*, State of California. Sacramento: California Environmental Protection Agency.

Changnon, S. A. (2005). Economic impacts of climate conditions in the United States: Past, present, and future, *Climatic Change, 68*, 1–9.

Clean Air Task Force (2004). *Wounded Waters: The Hidden Side of Power Plant Pollution*, Boston.

Considine, T. J. (2000). The impacts of weather variations on energy demand and carbon emissions, *Resource and Energy Economics*, *22*, 295-314.

Crichton, D. (2005). Insurance and Climate Change, in Conference on "Insurance and Climate Change," presented at Conference on "*Climate Change, Extreme Events, and Coastal Cities*," Houston, TX. Accessed at: http://64.233. 187.104/search e.edu/CentersandInst/ShellCenter/emplibrary/CoastalCities.pdf+Cichton,+coastal+cities,+2005&hl=en&gl=us&ct=clnk&cd=1

Cubasch, U. et al. (2001). Projections of Future Climate Change. In: *Climate Change 2001: The Scientific Basis. Contribution of Working Group I to the Third Assessment Report of the Intergovernmental Panel on Climate Change*, Cambridge: Cambridge University Press: 525-582.

Darmstadter, J. (1993). Climate change impacts on the energy sector and possible adjustments in the MINK region, *Climatic Change*, *24*, 117-129.

Daycock, C., DesJardins, R. & Fennell, S. (2004). Generation Cost Forecasting Using On-Line Thermodynamic Models. *Proceedings of Electric Power*, March 30-April 1, 2004, Baltimore, MD.

Dettinger, M. D. (2005). *Changes In Streamflow Timing In The Western United States In Recent Decades*, USGS Fact Sheet 2005-3018.

DOC/DOE (2001). *References to Deepwater Ports Act*. Accessed at: http://www.law10_29.html

DOE-2 (2006). *DOE-2 Building Energy Use and Cost Analysis Tool*. Accessed at: http://doe2.com/DOE2/index.html

DOI (2003). *Water 2025 – Preventing Crises and Conflict in the West*, Washington: U.S. Department of the Interior.

Downton, M. W. et al. (1988). Estimating historical heating and cooling needs: Per capita degree-days, *Journal of Applied Meteorology*, *27(1)*, 84–90.

Edmonds, J. et al. (1996a). An integrated assessment of climate change and the accelerated introduction of advanced energy technologies: An application of

Minicam 1.0, *Mitigation and Adaptation Strategies for Global Change 1(4)*, 311-339.

Edwards, A. (1991). *Global Warming From An Energy Perspective, Global Climate Change And California, Berkeley*, CA: University of California Press: Chapter 8.

Elkhafif, M. (1996). An iterative approach for weather-correcting energy consumption data, *Energy Economics 18(3)*, 221–23 0.

EIA (2001a). *Residential Energy Consumption Survey 2001: Consumption and Expenditure Data Tables.* Washington, DC: Energy Information Administration. Accessed at: http://www.eia.doe.gov/emeu/recs/recs 2001/detailcetbls.html

EIA (2001b). *2001 Public Use Data Files (ASCII Format)*, Washington, DC: Energy Information Administration. Accessed at: http://www.eia.doe.gov/ emeu/recs/recs2001/publicuse2001.html.

EIA (2001c). *U.S. Climate Zones*, Washington, DC: Energy Information Administration. Accessed at: http://www.eia.doe.gov/emeu/recs/ climate_zone.html.

EIA (2002). *Annual Coal Report*, DOE/EIA-0584, Washington, DC: Energy Information Administration.

EIA (2002a). Energy Consumed as a Fuel by End Use: Table 5.2 - By Manufacturing Industry with Net Electricity (trillion Btu), *2002 Energy Consumption by Manufacturers--Data Tables*, Washington, DC: Energy Information Administration. Accessed at: http://www.eia.doe.gov/emeu/ mecs/mecs2002/data02/shelltables. html.

EIA (2003). CBECS Public Use Microdata Files, Washington, DC: Energy Information Administration. Accessed at: http://www.eia.doe.gov/emeu/ cbecs/cbecs2003/public_use_2003/c becs_pudata2003.html

EIA (2004a). *Annual Energy Review*, Washington, DC: Energy Information Administration.

EIA (2004b). *Petroleum Supply Annual*, Washington, DC: Energy Information Administration.

EIA (2005d). Renewable Energy Trends 2004, Table 18, *Renewable Electric Power Sector Net Generation by Energy Source and State*, 2003. Washington, DC: Energy Information Administration.

EIA (2006). *Annual Energy Outlook 2006, with Projections to 2030.* DOE/EIA-0383(2006). Washington, DC: Energy Information Administration,

Elliott D. B. et al. (2004). *Methodological Framework for Analysis of Buildings-Related Programs: GPRA Metrics Effor*t. Pacific Northwest National Laboratory, PNNL-14697, Richland, WA.

EPA (2000). *Economic and Engineering Analyses of the Proposed Section 316(b) New Facility Rule*, Washington, DC: Environmental Protection Agency, EPA-821-R-00-019.

EPRI (2003). *A Survey of Water Use and Sustainability in the United States With a Focus on Power Generation*, Washington, DC: Electric Power Research Institute, EPRI Report No. 1005474.

Fankhauser, S. (1995). *Valuing Climate Change: The Economics of the Greenhouse*, London: Earthscan Publications, Ltd.

Fidje A. & Martinsen, T. (2006). *Effects of Climate Change on the Utilization of Solar Cells in the Nordic Region.* Extended abstract for European Conference on Impacts of Climate Change on Renewable Energy Sources. Reykjavik, Iceland, June 5-9, 2006.

Flavin, C. & Lenssen, N. (1994). *Power Surge: Guide to the Coming Energy Revolution*, New York: WW Norton.

Franco, G. (2005). *Climate Change Impacts and Adaptation in California*, CEC-500-2005-103-SD, California Energy Commission, Sacramento, CA.

Franco, G. & Sanstad, A. (2006). *Climate Change and Electricity Demand in California*, Final white paper, California Climate Change Center, CEC-500-2005-201-SD. Accessed at: http://www.climatechange.ca.gov/climate_action_team/reports/index.html

GAO (2003). *Freshwater Supply, States' Views of How Federal Agencies Could Help Them Meet the Challenges of Expected Shortages*, Washington, DC: Government Accountability Office.

GAO (2004). *Climate Change: Analysis of Two Studies of Estimated Costs of Implementing the Kyoto Protocol.* Washington, DC: Government Accountability Office.

Gellings, C. & Yeager, K. (2004). "Transforming the electric infrastructure," *Physics Today*, *57*, 45-51.

Georgakakos, K. et al. (2005). Integrating climate-hydrology forecasts and multi-objective reservoir management for Northern California. *Earth Observing Systems (EOS)*, *86(12)*, 122-127.

Greenwire (2003). *State orders N.Y.'s Indian Point to take steps to protect fish*, Greenwire. Accessed at: http://www.eenews.net/Greenwire.htm.

Hadley, S. W. et al. (2004). *Future U.S. Energy Use for 2000-2 025 as Computed with Temperature from a Global Climate Prediction Model and Energy Demand Model* 24th USAEE/IAEE North American Conference, Washington, DC.

Hadley, S. W. et al. (2006). Responses of energy use to climate change: A climate modeling study, *Geophysical Research Letters 33*, L17703, doi: 10.1029/2006GL026652, 2006.

Harrison, G. & Wallace, A. (2005). Climate sensitivity of marine energy, *Renewable Energy*, *30*, 1801–18 17.

Hayhoe, K. et al. (2004). Emissions pathways, climate change, and impacts on California. P*roceedings, National Academy of Sciences (NAS), 101/34*, 12422-12427.

Hill, D. & Goldberg, R. (2001). Energy Demand. In: C. Rosenzweig, and W. Solecki, (eds.), *Climate Change and a Global City: An Assessment of the Metropolitan East Coast Region*, Columbia Earth Institute, New York: 121-147.

Hoffert, M. I. et al. (2002). Advanced technology paths to global climate stability: Energy for a greenhouse planet, *Science, 298*, 98 1-87.

Huang, Y. J. (2006). *The Impact of Climate Change on the Energy Use of the U.S. Residential and Commercial Building Sectors*, LBNL60754, Lawrence Berkeley National Laboratory, Berkeley CA.

Johnson, V. H. (2002). *Fuel Used for Vehicle Air Conditioning: A State-byState Thermal Comfort-Based Approach*, SAE paper number 2002-01-1957. Future Car Congress, June 2002, Hyatt Crystal City, VA, USA, Session: Climate Control Technology. SAE International, Warrendale, PA 15096-0001. Accessed at: http://www.sae.org/technical/papers/2002-01-1957

ICF (1995). *Potential Effects of Climate Change on Electric Utilities, TR105005, Research Project 2141-11,* Prepared for Central Research Institute of the Electric Power Industry (CRIEPI) and the Electric Power Research Institute (EPRI).

IEA (2004). *Energy Security and Climate Change Policy Interactions*. Information Paper, International Energy Agency, Paris.

IPCC (2001). *Climate Change, 2001: Mitigation*. Contribution of Working Group III to the Third Assessment Report of the Intergovernmental Panel on Climate Change. Cambridge: Cambridge University Press.

IPCC (2001a). *Climate Change 2001: Impacts, Adaptation and Vulnerability.*, Cambridge: Cambridge University Press.

IPCC (2005a). *Special Report on Carbon Dioxide Capture and Storage, Contribution of Working Group III to the Third Assessment Report of the Intergovernmental Panel on Climate Change.* Cambridge: Cambridge University Press.

IPPC (2005b). *Workshop on New Emission Scenarios*, Working Group III Technical Support Unit, 29 June – 1 July 2005, Laxenburg, Austria. Accessed at: http://www.ipcc.ch/meet/othercorres/ESWmeetingreport.pdf

IPCC (2007). *Climate change 2007: Mitigation*. Contribution of Working Group III to the Fourth Assessment Report of the Intergovernmental Panel on Climate Change [B. Metz, O. R. Davison, P. R. Bosch, R. Dave, and L. A. Meyer (eds)], Cambridge: Cambridge University Press.

Interlaboratory Working Group (2000). *Scenarios for a Clean Energy Future*. Prepared by Lawrence Berkeley National Laboratory (LBNL-44029) and Oak Ridge National Laboratory (ORNL/CON476) for the U.S. Department of Energy.

Jorgenson, D. et al. (2004). *U.S. Market Consequences Of Global Climate Change*, Pew Center on Global Climate Change, Washington, DC

Kainuma, M. et al. (2004). Analysis of global warming stabilization scenarios: the Asian-Pacific integrated model, *Energy Economics*, *26*, 709– 719.

Kelly et al. (2005). *Emissions Trading: Developing Frameworks and Mechanisms for Implementing and Managing Greenhouse Gas Emissions*," Proceedings of the World Energy Engineering Congress, September 14-16, 2005, Austin, TX.

Kurosawa, A. (2004). Carbon concentration target and technological choices, *Energy Economics*, *26*, 675– 684.

Lam, J. C. (1998). Climatic and economic influences on residential electricity consumption, *Energy Conversion Management*, *39(7)*: 623–629.

Land Letter (2004). *Western Power Plants Come Under Scrutiny as Demand and Drought Besiege Supplies*. Accessed at: http://www.eenews.net/Landletter.htm.

Le Comte, D. M. & Warren, H. E. (1981). Modeling the impact of summer temperatures on national electricity consumption, *Journal of Applied Meteorology*, *20*, 1415–1419.

Lehman, R. L. (1994). Projecting monthly natural gas sales for space heating using a monthly updated model and degree-days from monthly outlooks, *Journal of Applied Meteorology, 33(1)*, 96–106.

Lettenmaier, D. P. et al. (1999). Water resources implications of global warming: A U.S. regional perspective, *Climate Change 43(3)*, 537-579.

Linder, K. P. & Inglis, M. R. (1989). *The Potential Impact of Climate Change on Electric Utilities, Regional and National Estimates*, Washington, DC: U.S. Environmental Protection Agency.

London Climate Change Partnership (2002). London's Warming, London: UK *Climate Impacts Programme*.

Loveland, J. E. & Brown, G. Z. (1990). *Impacts of Climate Change on the Energy Performance of Buildings in the United States*, U.S. Congress, Office of Technology Assessment, Washington, DC, OTA/UW/UO, Contract J3-4825.0.

Mansur, E. T., Mendelsohn, R. & Morrison, W. (2005). A discrete-continuous choice model of climate change impacts on energy, SSRN Yale SOM Working Paper No. ES-43 (abstract number 738544), Submitted to *Journal of Environmental Economics and Management.*

Marketwatch.com (2006). June 9, 2006, Available at: http://www.marketwatch.com/

Maulbetsch, J. S. & DiFilippo, M. N. (2006). *Cost and Value of Water Use at Combined Cycle Power Plants*, California Energy Commission, PIER Energy-Related Environmental Research, CEC-500-2006-034, April 2006.

Mendelsohn, R., Nordhaus, W. & Shaw, D. (1994). The impact of global warming on Agriculture: A Ricardian Analysis," *American Economic Review, 84*, 753-771.

Mendelsohn, R. & Neumann, J. eds. (1999). *The Economic Impact of Climate Change on the Economy of the United States*, Cambridge: Cambridge University Press.

Mendelsohn, R., Morrison, W., Schlesinger, M. & Andronova, N. (2000). Country-specific market impacts from climate change, *Climatic Change 45*, 553-569

Mendelsohn, R. (ed.) (2001). *Global Warming and the American Economy: A Regional Assessment of Climate Change*, Cheltenham Glos, UK: Edward Elgar Publishing,.

Mendelsohn, R. (2003). 'The Impact of Climate Change on Energy Expenditures in California.' Appendix XI in Wilson, T., and L. Williams, J. Smith, R. Mendelsohn, *Global Climate Change and California: Potential Implications for Ecosystems, Health, and the Economy*, Consultant report 500-03-058CF to the Public Interest Energy Research Program, California Energy Commission, August 2003. Available from http://www.energy.

Mendelsohn, R. & Williams, L. (2004). *Comparing Forecasts of the Global Impacts of Climate Change*" Mitigation and Adaptation Strategies for Global Change, 9 (2004), 315-333.

Meyer, J. L. et al. (1999). Impacts of climate change on aquatic ecosystem functioning and health, *J. Amer. Water Resources Assoc. 35(6)*, 1373-1386.

Miller, B. A. & Brock, W. G. (1988). *Sensitivity of the Tennessee Valley Authority Reservoir System to Global Climate Change*, Report No. WR28-1-680-101. Tennessee Valley Authority Engineering Laboratory, Norris, TN.

Miller, N. L. et al. (2007). Climate, extreme heat and energy demand in California, *Journal of Applied Meteorology and Climatology*, in press.

Milwaukee Journal Sentinel (2005). *Wisconsin Energy Just Can't Stay Out of the News with Their Intake Structures*, February 18, 2005.

Morris, M. (1999). *The Impact of Temperature Trends on Short-Term Energy Demand*, Washington, DC: Energy Information Administration.

Morrison, W. N. & Mendelsohn, R. (1999). The impact of global warming on U.S. energy expenditures. In: [R. Mendelsohn and J. Neumann, (eds.)], *The Economic Impact of Climate Change on the United States Economy*, Cambridge: Cambridge University Press: 209–236.

NACC (2001). *Climate Change Impacts on the United States: The Potential Consequences of Climate Variability and Change*, Washington: U.S. Global Change Research Program.

Nakicenovic, N. & Riahi, K. (2003). *Model Runs With MESSAGE in the Context of the Further Development of the Kyoto-Protocol.* Laxenburg: IIASA. Available from: http://www.wbgu.de/wbgu_sn2003_ex03.pdf

NARUC (2006). "*Enhancing the Nation's Electricity Delivery System - Transmission Needs,*" Mark Lynch, 2006 National Electricity Delivery Forum, Washington, DC, February 15, 2006. Accessed at: http://www.energetics.com/electricity

NASEO (2005). *Florida State's Energy Emergency Response to the 2004 Hurricanes, National Association of State Energy Officials for the Office of Electricity Delivery and Energy Reliability*, Department of Energy, June 2005. Available at: http://www.naseo.org/Committees/energysecurity/documents/florida_response.pdf

Nash, L. L. & Gleick, P. H. (1993). *The Colorado River Basin and Climate Change: The Sensitivity of Stream Flow and Water Supply to Variations in Temperature and Precipitation*, EPA230-R-93-009, Washington, DC: U.S. Environmental Protection Agency,

National Commission on Energy Policy (2004). *Ending the Energy Stalemate: A Bipartisan Strategy to Meet America's Energy Challenges*. Washington, DC: National Commission on Energy Policy (NCEP),

Niemeyer, V. (2005). *Climate science needs for long-term power sector investment decisions*, Presented at the CCSP Workshop on Climate Science in Support of Decision Making, Washington, DC. November 15, 2005.

Nordhaus, W. D. (2006). Geography and Macroeconomics: New Data and New Findings. *Proceedings of the National Academy of Sciences*. Accessed at: http://www.pnas.org/cgi/doi/10.1073/nas.0509842103

Nordhaus, W. D. and J. G. Boyer, 2000: *Warming the World: Economic Models of Global Warming*, Cambridge: MIT Press.

Northwest Power and Conservation Council (2005). *Effects of Climate Change on the Hydroelectric System,* Appendix N, The Fifth Northwest Electric Power and Conservation Plan. Document 2005-7. Northwest Power and Conservation Council, Portland, Oregon. Accessed at: http://www.nwcouncil.org/energy/

NRC (2002). *Abrupt Climate Change: Inevitable Surprises*, National Research Council, Washington, DC: National Academy Press.

NREL (2006). "On The Road To Future Fuels And Vehicles," *Research Review.* May 2006, NREL/BR-840-3 8668.

O'Keefe, W. (2005). *Climate Change and National Security*, Marshall Institute, May 2005. Accessed at: http://www.marshall.org/article.php?id

Ohmura, A. & Wild, M. (2002). Is the hydrologic cycle accelerating? *Science 298*, 1345-1346.

O'Neill, R. (2003). "*Transmission Investments and Markets, Federal Energy Regulation Commission*," presented at Harvard Electricity Policy Group meeting, Point Clear, AL, December 11, 2003. Available from: http://www.ksg.harvard.edu/hepg/Papers/Oneill.trans. markets.1 1.Dec.03.pdf

ORNL (2006). See the Oak Ridge National Laboratory Bioenergy Feedstock Information Network, http://bioenergy.ornl.gov/main.aspx, and the Agricultural Research Service Bioenergy and Energy Alternatives Program, see: http://www.ars.usda.gov/research/programs/programs.htm?np_code=307, for examples of the extensive research in this area.

Ouranos (2004). *Adapting to Climate Change.* Montreal, Canada, Ouranos Consortium,.

Pacala, S. & Socolow, R. (2004). Stabilization wedges: Solving the climate problem for the next 50 years with current technologies, *Science*, *305*, 968-72.

Paltsev, S. et al. (2005). *The MIT Emissions Prediction and Policy Analysis (EPPA) Model: Version 4*, MIT Global Change Joint Program Report Series #125

Pan, Z. T. et al. (2004). On the potential change in solar radiation over the U.S. due to increases of atmospheric greenhouse gases, *Renewable Energy*, *29*, 1923-1928.

Pardo, A., Meneu, V. & Valor, E. (2002). Temperature and seasonality influences on the Spanish electricity load, *Energy Economics 24(1)*, 55–70.

Parker, L. S. (1999). *Electric Utility Restructuring: Overview Of Basic Policy Questions*, Washington, DC: Congressional Research Service Report 97-154.

Parker, D. S. (2005). *Energy Efficient Transportation for Florida*, Florida Solar Energy Center, University of Central Florida, Cocoa, Florida, Energy Note FSEC-EN-19. Accessed at: http://www.fsec.ucf.edu/Pubs/energynotes/en-19.htm.

Parson, E. et al. (2003). Understanding climatic impacts, vulnerabilities, and adaptation in the United States: Building a capacity for assessment, *Climatic Change 57*, 9–42.

Petroleum Institute for Continuing Education, undated: *Fundamentals of Petroleum Refining Economics, training course curriculum.*

Pipeline Engineering (2007). *Pipeline facts and history*. Accessed at: http://alyeska-pipe.com/Pipelinefacts/PipelineEngineering.html

Placet, M., Humphreys, K. K. & Mahasenan, N. (2004). *Climate Change Technology Scenarios: Energy, Emissions, And Economic Implications*, Pacific Northwest National Laboratory Richland, WA, PNNL-14800. Available from: http://www.pnl.gov/energy/

PNNL (2002). *Facility Energy Decision System User's Guide, Release 5.0*, PNNL-10542 Rev. 3. Prepared for the U.S. Department of Energy - Federal Energy Management Program, U.S. Army - Construction Engineering Research Laboratory, U.S. Army - Forces Command, Defense Commissary Agency, and U.S. Navy - Naval Facilities Engineering Service Center, Pacific Northwest National Laboratory, Richland, WA.

Quayle, R. G. & Diaz, H. F. (1979). Heating degree-day data applied to residential heating energy consumption, *Journal of Applied Meteorology 19*, 241–246

Reno-Gazette Journal 2005: *Nevada Residents Wary of Sempra Water Rights Purchases*, February 22, 2005.

Rose, A., Kamat, R. & Abler, D. (1999). The economic impact of a carbon tax on the Susquehanna River Basin economy, *Energy Economics*, *21*, 363-84.

Rose, A. & Oladosu, G. (2002). Greenhouse gas reduction in the U.S.: Identifying winners and losers in an expanded permit trading system, *Energy Journal*, *23*, 1-18.

Rose, A. & Zhang, Z. (2004). Interregional burden-sharing of greenhouse gas mitigation in the United States," *Mitigation and Adaptation Strategies for Global Change*, *9*, 477-500.

Rose, A. & Oladosu, G. (2006). Income distribution impacts of climate change mitigation policy in the Susquehanna River Basin," *Energy Economics, 29(3)*, 520-544.

Rosenthal, D. H., Gruenspecht, H. K. & Moran, E. (1995). Effects of global warming on energy use for space heating and cooling in the United States, *Energy Journal 16(2)*, 77-96.

Rosenzweig, C. & Solecki, W. D. (eds.) (2001). *Climate Change and a Global City: The Potential Consequences of Climate Variability and Change – Metro East Coast (MEC).* Report for the U.S. Global Change Research Program, National Assessment of Possible Consequences of Climate Variability and Change for the United States, New York: Columbia Earth Institute.

Ruosteenoja, K. et al. (2003). Future Climate in World Regions: An Intercomparison of Model-Based Projections for the New IPCC Emissions Scenarios, *The Finnish Environment 644,* Helsinki, Finland: Finnish Environment Institute.

Ruth, M. & Lin, A-C. (2006). Regional energy and adaptations to climate change: Methodology and application to the state of Maryland, *Energy Policy, 34*, 2820-2833.

Rutherford, T. (2001). *Equity and Global Climate Change: Economic Considerations*, Discussion Brief Prepared for the Pew Center for Global Climate Change - Equity and Global Climate Change Conference, Washington, DC, April, 2001.

Sailor, D. J. (2001). Relating residential and commercial sector electricity loads to climate: Evaluating state level sensitivities and vulnerabilities, *Energy 26(7)*, 645–657.

Sailor, D. J. & Muñoz, J. R. (1997). Sensitivity of electricity and natural gas consumption to climate in the U.S: Methodology and results for eight states, *Energy, 22(10)*, 987–998.

Sailor, D. J. & Pavlova, A. A. (2003). Air conditioning market saturation and long-term response of residential cooling energy demand to climate change, *Energy, 28(9)*, 941–951.

Schwarz, P. & Randall, D. (2004). *Abrupt Climate Change*, Washington: GBN. Available at: http://www.gbn.com/ArticleDisplayServlet.srv?aid 1.

Scott, M. J. et al. (1993). *The Effects of Climate Change on Pacific Northwest Water-Related Resources: Summary of Preliminary Findings*, Pacific Northwest Laboratory, PNL-8987, Richland, Washington.

Scott, M. J., Hadley, D. L. & Wrench, L. E. (1994: Effects of climate change on commercial building energy demand, *Energy Sources, 16(3)*, 339–354.

Scott, M. J., Dirks, J. A. & Cort, K. A. (2005). The adaptive value of energy efficiency programs in a warmer world: Building energy efficiency offsets effects of climate change, PNNL-SA-451 18. In: *Reducing Uncertainty Through Evaluation*, Proceedings of the 2005 International Energy Program Evaluation Conference, August 17- 19, 2005, Brooklyn, New York.

Segal, M., Pan, Z., Arritt, R. W. & Takle, E. S. (2001). On the potential change in wind power over the U. S. due to increases of atmospheric greenhouse gases, *Renewable Energy, 24*, 235-243.

Senate of Texas (1999). *Interim Committee Report*, Interim Committee on Electric Utility Restructuring, Austin, Texas

Shurepower, LLC (2005). *Electric Powered Trailer Refrigeration Unit Market Study and Technology Assessment*, Agreement 8485-1, June 24, 2005. Prepared for New York State Energy Research and Development Authority. Rome, New York: Shurepower, LLC.

SNL (2006a). *Energy and Water Research Directions - A Vision for a Reliable Energy Future*, Sandia National Laboratories, Albuquerque, NM.

SNL (2006b). *Energy and Water Research Directions – A Vision for a Reliable Energy Future*, Sandia National Laboratories, Albuquerque, NM

Stress Subsea, Inc. (2005). *Deep Water Response to Undersea Pipeline Emergencies*. Final Report, Document No. 221006-PL-TR-0001, Houston, TX.

Struck, D. (2006). Canada pays environmentally for U.S. oil thirst: Huge mines rapidly draining rivers, cutting into forests, boosting emissions, W*ashington Post*, May 31, 2006, A01.

Subcommittee on Conservation, Credit, Rural Development, and Research, Of The Committee on Agriculture, House of Representatives. (2001). *Energy Issues Affecting the Agricultural Sector of the U.S. Econo*my. One Hundred Seventh Congress, First Session, April 25 and May 2, 2001. Serial No. 107–6. Printed for the Use of the Committee on Agriculture. Washington, DC: U.S. Government Printing Office.

Sweeney, J. L. (2002). *The California Electricity Crisis*. Pub. No. 503, Hoover Institution Press, Stanford, California.

Tol, R. S. J. (2002). Welfare specifications and optimal control of climate change: An application of fund, *Energy Economics, 24*, 367-376.

Union of Concerned Scientists (1999). *Confronting Climate Change in California, with the Ecological Society of America*, Cambridge, Massachusetts

University of Georgia College of Agricultural and Environmental Sciences et al. (2005). Georgia Annual Report of Accomplishments FY 2004. *Agricultural Research and Cooperative Extension Programs*, University of Georgia & Fort Valley State University, Athens, Georgia, 100 pp.

University of Missouri-Columbia (2004). *Influence of Missouri River on Power Plants and Commodity Crop Prices*, Food and Agriculture Policy Institute.

U.S. Arctic Research Commission (2003). *Climate Change, Permafrost, and Impacts on Civil Infrastructure*, Permafrost Task Force Report, p. 10.

USGS (2000). *National Assessment of Coastal Vulnerability to Future Sea-Level Rise*, USGS Fact Sheet FS-076-00, Washington, DC: U.S. Geological Survey.

USGS (2004). Estimated Use of Water in the United States in 2000, *USGS Circular 1268*, Washington, DC: U.S. Geological Survey.

U.S. Climate Change Technology Program (CCTP) (2005). *2005 Strategic Plan: Draft for Public Comment*, September 2005.

Van Vuuren, D. P., de Vries, B., Eickhout, B. & Kram, T. (2004). Responses to technology and taxes in a simulated world, *Energy Economics, 26*, 579– 601.

Van Vuuren, D. P.,. Weyant, J. & de la Chesnaye, F. (2006). Multi-gas scenarios to stabilize radiative forcing, *Energy Economics, 28*, 102–120

Vicuña, S., Leonardson, R. & Dracup, J. A. (2006). *Climate Change Impacts On High-Elevation Hydropower Generation In California's Sierra Nevada: A Case Study In The Upper American River*, CEC-500-2005-199-SF, California Energy Commission, Sacramento, California. Available at: http://www.climatechange.ca.gov/climate_action_team/reports/ind ex.html

Warren, H. E. & LeDuc, S. K. (1981). Impact of climate on energy sector in economic analysis, *Journal of Applied Meteorology, 20*, 1431– 1439.

Westenburg, C.L., DeMeo G.A., and Tanko, D.J., 2006: Evaporation from Lake Mead, Arizona and Nevada, 1997–99: U.S. Geological Survey Scientific Investigations Report 2006-5252, 24 p.

Weyant, J. P., de la Chesnaye, F. C. & Blanford, G. J. (2006). Overview of EMF-21: Multigas mitigation and climate policy, *The Energy Journal, 27* (Multi-Greenhouse Gas Mitigation and Climate Policy, Special Issue #3), 1-32.

Wilbanks, T. J. (2005). Issues in developing a capacity for integrated analysis of mitigation and adaptation, *Environmental Science & Policy, 8*, 541–547.

Wilbanks, T. et al. (2007). Toward an integrated analysis of mitigation and adaptation: some preliminary findings. In: T.Wilbanks, J. Sathaye, and R. Klein, (eds.), "Challenges in Integrating Mitigation and Adaptation as Responses to Climate Change," special issue, *Mitigation and Adaptation Strategies for Global Change, 12*,713-725.

Winkelmann, F. C. et al. (1993). *DOE-2 Supplement Version 2*. 1E, LBL-34947, Lawrence Berkeley National Laboratory, Berkeley CA.

WWF (2003). *Power switch: impacts of climate policy on the global power sector*, Washington, DC: WWF International.

Yan, Y. Y. (1998). Climate and residential electricity consumption in Hong Kong, *Energy, 23(1)*, 17–20.

In: Effects of Climate Change on Energy Production... ISBN: 978-1-60741-426-1
Editor: Alrik M. Solberg

Chapter 8

ANNEX A – TECHNICAL NOTE: METHODS FOR ESTIMATING ENERGY CONSUMPTION

U.S. Climate Change Science Program

Previous authors have used a number of approaches to estimate the impact of climate change on energy use in U.S. buildings. Many of the researchers translate changes in average temperature change on a daily, seasonal, or annual basis into heating and cooling degree days, which are then used in building energy simulation models to project demand for space heating and space cooling (e.g., Rosenthal et al. 1995, Belzer et al. 1996, and Amato et al. 2005). Building energy simulation is often done directly with average climate changes used to modify daily temperature profiles at modeled locations (Scott et al. 2005, and Huang 2006). (See Box A. 1 on heating and cooling degree-days.)

Building energy simulation models such as CALPAS3 (Atkinson et al. 1981), DOE-2 (Winkelmann et al. 1993), or FEDS and BEAMS (PNNL 2002, Elliott et al. 2004) have been used to analyze the impact of climate warming on the demand for energy in individual commercial buildings only (Scott et al. 1994) and in groups of commercial and residential buildings in a variety of locations (Loveland and Brown 1990, Rosenthal et al. 1995, Scott et al. 2005, and Huang 2006).

Other researchers have used econometrics and statistical analysis techniques (most notably the various Mendelsohn papers discussed in Chapter 2, but also the Belzer et al. 1996 study using the CBECS micro- data, and Sailor and Muñoz 1997, Sailor 2001, Amato et al. 2005, Ruth and Lin 2006, and Franco and Sanstad

2006, using various state-level time series.) A subcategory of the econometric technique is cross-sectional analysis. For example, Mendelsohn performed cross-sectional econometric analysis of the RECS and CBECS microdata sets to determine how energy use in the residential and commercial building stock relates to climate (Morrison and Mendelsohn 1999; Mendelsohn 2001), and then used the resulting equations to estimate the future impact of warmer temperatures on energy consumption in residential and commercial buildings. Mendelsohn 2003 and Mansur et al. 2005 subsequently elaborated the approach into a complete and separate set of discrete-continuous choice models of energy demand in residential and commercial buildings.

Box A.1 Heating and Cooling Degree-Days and Building Energy Use

Energy analysts often refer to concepts called heating and cooling degree-days when calculating the impact of outdoor temperature on energy use in buildings. Buildings are considered to have a minimum energy use temperature where the building is neither heated nor cooled, and all energy use is considered to be nonclimate sensitive. This is called the "balance point" for the building. Each degree deviation from that balance point temperature results in heating (if the temperature is below the balance point) or cooling (if the temperature is above the balance point). For example, if the balance point for a building is 60°F and the average outdoor temperature for a 30-d period is 55°F, then there are 5 x 30 heating degree days for that period. Energy demand is usually considered to increase or decrease proportionately with increases in either heating degree-days or cooling degree-days.

Balance points by default are usually considered to be 65°F because many weather datasets come with degree-days already computed on that basis (See Amato et al 2005). However, empirical research on regional datasets and on the RECS and CBECS microdata sets suggests that regional variations are common. In Massachusetts, for example, Amato et al. found a balance point temperature for electricity in the residential sector of 60°F and 55°F for the residential sector. Belzer et al. (1996) found that the newer commercial buildings have even lower balance point temperatures, probably because of tighter construction and the dominance of lighting and other interior loads that both aid with heating and make cooling more of a challenge.

Finally, Hadley et al. 2004, 2006, directly incorporated changes in heating degree-days and cooling degree- days expected as a result of climate change into the residential and commercial building modules of the Energy Information Administration's National Energy Modeling System, so that their results incorporated U.S. demographic trends, changes in building stock and energy-using equipment, and (at least some) consumer reactions to energy prices and climate at a regional level. Hadley et al. translated temperatures from a single climate scenario of the Parallel Climate Model into changes in heating degree days (HDDs) and cooling degree-days (CDDs) that are population-averaged in each of the nine U.S. Census divisions (on a 65° F base –against the findings of Rosenthal et al., Belzer et al., and Mansur et al. 2005, all of which projected a lower balance point temperature for cooling and a variation in the balance point across the country). They then compared these values with 1971- 2000 average HDDs and CDDs from the National Climate Data Center for the same regions. The changes in HDD and CDD were then used to drive changes in a special version (DD-NEMS) of the National Energy Modeling System (NEMS) of the U.S. Energy Information Administration, generally used to provide official energy consumption forecasts for the Annual Energy Outlook (EIA 2006). Table A. 1 contains a summary of methods used in the various studies employed in this chapter.

Table A.1 Methods Used in U.S. Studies of the Effects of Climate Change on Engergy Demand in Buildings

Authors	Methods	Comments
National Studies		
Linder-Inglis 1989	Electric utility planning model	Electricity only. Results available for 47 state and substate service areas. Calculates peak demand.
Rosenthal et al.1995	Reanalysis of building energy consumption in EIA Annual Energy Outlook	Energy-weighted national averages of census division-level data
Belzer et al. 1996	Econometrics on CBECS commercial sector microdata	Used HDD and CDD and estimated energy balance points

Table A.1. (Continued)

Authors	Methods	Comments
Mendelsohn 2001	Econometric analysis of RECS and CBECS microdata	Takes into account energy price forecasts, market penetrateion of air conditioning. Precipitation increases 7%.
Scott et al. 2005	Building models (FEDS and BEAMS)	Varies by region. Allows for growth in residential and commercial building stock, but not increased adoption of air conditioning in response to warming
Mansur et al. 2005	Econometric analysis of RECS and CBECS microdata	Takes into account energy price forecasts, market penetration of air conditioning. Precipitation increases 7%. Affects both fuel choice and use.
Hadley et al. 2004; 2006	NEMS energy model, modified for changes in degree-days	Primary energy, residential and commercial combined. Allows for growth in residential and commercial building stock.
Huang et al. 2006	DOE-2 building energy model	Impacts vary by region, building type.
Regional Studies		
Loveland and Brown 1990	CALPAS3 Building Energy Model	Single family detached house, commercial building, 6 individual cities
Baxter and Calandri 1992	Building energy model	Electricity only, California.
Scott et al. 1994	DOE-2 building energy model	Small office building, 4 specific cities
Sailor 2001	Econometric on state time series	Total electricity per capita in 7 out of 8 energy-intensive states; one state (Washington) used electricity for space heating
Sailor and Pavlova 2003	Econometric on state-level time series	Four states. Includes increased market saturation of air conditioning
Mendelsohn 2003	Econometric on national cross sectional data on RECS and CBECS data	Impacts for California only. Residential and commercial. Expenditures on energy.

Amato et al. 2005	Time series econometric on state data	Massachusetts (North), Winter monthly residential capita consumption, commercial monthly per employee consumption
Ruth and Lin 2006	Time series econometric on state data	Maryland (borderline North-South), residential natural gas, heating oil, electricity expenditures
Franco and Sanstad 2006	Regression of electricity demand in California Independent System Operator with average daily temperature and daily consumption in the CalISO area in 2004, and the relationship between peak demand and average daily maxi-mum temperature over the period 1961–1990	Electricity only

In: Effects of Climate Change on Energy Production… ISBN: 978-1-60741-426-1
Editor: Alrik M. Solberg

Chapter 9

Annex B – Organizations and Individuals Consulted

U.S. Climate Change Science Program

Vicki Arroyo	Pew Center
Malcolm Asadoorian	Massachusetts Institute of Technology
Kelly Birkinshaw	California Energy Commission
Benjamin J. DeAngelo	U.S. Environmental Protection Agency
Richard Eckaus	Massachusetts Institute of Technology
Bill Fang	Edison Electric Institute
Guido Franco	California Energy Commission
Howard Gruenspecht	Energy Information Administration
Jay Gullege	Pew Center
Richard C Haut	Houston Advanced Research Center
Haroon S. Kheshgi	ExxonMobil Research and Engineering Co.
Joe Loper	Alliance to Save Energy
Sasha Mackler	National Commission on Energy Policy
Victor Niemeyer	Electric Power Research Institute
Paul Pike	Ameren Corporation
Will Polen	U.S. Energy Association
Walt Retzch	American Petroleum Institute
Terry Surles	University of Hawaii
Tom Wilson	Electric Power Research Institute

Laurie ten Hope	California Energy Commission
Barry Worthington	U.S. Energy Association
Kate Zyla	Pew Center

In: Effects of Climate Change on Energy Production... ISBN: 978-1-60741-426-1
Editor: Alrik M. Solberg

Chapter 10

ANNEX C – GLOSSARY

U.S. Climate Change Science Program

A

Adaptation In climate change discussions, refers to actions that respond to climate change risks and/or impacts by reducing sensitivity to climate variables and/or increasing coping capacity

Aerosols A substance packaged under pressure with a gaseous propellant for release as a spray of fine particles

Ambient temperatures The temperature of the air surrounding a power supply or heating/cooling medium

Analytic-deliberative practices Combining systematic analysis with processes for collective qualitative consideration of broader issues

Aquifer An underground bed or layer of earth, gravel, or porous stone that yields water

B

Biodiesel An oxygenated fuel, primarily alkyl (methyl or ethyl) esters, produced from a range of biomass-derived feedstocks including oilseeds, waste vegetable oils, cooking oil, animal fats and trap grease, which can be used in blends or in "neat" form in compression-ignition engines to reduce emissions and improve engine performance

Btus British thermal units, a quantity of energy

Building equipment Energy-using equipment within a building, such as electric appliances

Building shell The external envelope of a building, including foundation, floor, walls, windows, outside doors, and roof

Building stock The total quantity of buildings in an area or sector of interest

C

Canadian model A climate change projection model from the Canadian Climate Change Centre (CGCM1), used in the U.S. National Assessment of Possible Consequences of Climate Variability and Change (2001)

Cap-and-trade A market-based system of limiting emissions in which a limited number of emissions permits are issued in the aggregate (cap); these permits are then freely exchangeable in markets (trade)

Cellulosic Pertaining to cellulose, a constituent of plant tissues and fibers

Climate change Changes in climate that depart from normal variability, representing significant changes in averages and/or extremes

Climate change impacts Effects of climate changes such as temperature change, precipitation change, severe weather events, and sea level rise on human and/or natural systems

Climate change related policies Public policy interventions in response to concerns about or impacts of climate change

Climate forcing effect Increases in certain trace gas molecules in the atmosphere that change the balance between incoming solar radiation and re-radiation of energy into space, leading to long-term atmospheric warming

Climate variability Changes in climate around averages, not necessarily associated with climate change

Climate-sensitive Refers to systems or phenomena whose behavior is noticeably affected by differences in climate

Closed-cycle cooling A method of cooling power plants in which water is withdrawn from a body of water, passed through the facility to cool power-production processes, cooled down in a cooling tower or similar method, and then reused for cooling

Combined cycle An electric-power generating method in which combustible gases are burned in a combustion turbine (topping cycle) and high-temperature gases from that operation are used to raise steam that is passed through a steam turbine (bottoming cycle). Both cycles drive electric generators

D

Delivery forms Forms in which energy is delivered to users: solid, liquid, gaseous, electricity

Demographic Related to the size, growth, and distribution of human populations

Discrete-continuous choice models A family of economic models in which the probability of a handful of choices (e.g., whether or not to select a particular heating technology) are modeled mathematically as a function of continuous variables such as income and price

Distribution systems Systems for moving energy delivery forms from producers to users

E

Econometric A field of economics that applies statistical procedures to mathematical models

Elasticities Refers to changes in one variable as the result of changes in another variable

Empirical Derived from observation or experiment, generally implying quantitative data

Energy consumption The amount of fuels and electricity (measured in common units such as British thermal units or Btus) utilized during a period of time to provide a useful service such as heating, cooling, or transportation

Energy conversion Changing energy-bearing substances from one form to another; e.g., petroleum refining or electric power generation

Energy demand The quantities of energy desired in the marketplace at various prices.

Energy infrastructure The capital equipment used to supply energy; e.g., power plants, refineries, natural gas pipelines, electric power lines and substations, etc.

Energy intensity The amount of energy consumed per unit of desired service

Energy markets Groups of buyers and sellers of energy goods and services and the institutions that make such exchanges possible

Energy prices Prices of petroleum and petroleum fuels, natural gas and manufactured gases, coal, uranium fuels, other fuels, and electricity, formed in energy markets via buying and selling processes

Energy production xtraction, conversion, and transportation of fuels and electricity to ultimate end use

Energy security Reliable and predictable supplies of fuels and electricity in national markets at stable prices, usually associated with the concerns about reliability of foreign supplies

Energy use See energy consumption

Ethanol An alcohol fuel produced chemically from ethylene or biologically from the fermentation of various sugars from carbohydrates found in agricultural crops and cellulosic residues from crops or wood. Often made from plants such as corn and typically blended in various proportions with conventional gasoline to make transportation fuel (gasohol)

Extreme weather events Weather events that are infrequent or unusual in their magnitude or intensity

F

Fossil fuels Hydrocarbon fuels derived from fossils: coal, petroleum, natural gas

Fuel types End-use delivery forms for energy: solid, liquid, gaseous, electricity

G

Gas turbine A rotary engine that extracts energy from a flow of combustion gas

Global Change Research Act of 1990 An act of the U.S. Congress that established the U.S. Global Change Research Program and called for periodic assessments of climate change implications for the U.S.

H

Hadley Centre Model A well-known British model for projecting climate change

Heating loads The amounts of energy necessary to keep the internal temperature in a building above a specific temperature range

Hydropower Hydroelectric power, derived from the energy value of running water

I

Indirect effects Effects derived not from the primary driver of interest but from effects of that driver on another system, process, or phenomenon

Integrated Resource Planning An approach to electric utility planning that integrates demand-side planning with supply-side planning

Intensity A measure of concentration, such as the amount of energy consumed for a particular purpose

K

Knowledge base The stock of knowledge about a particular topi

kWh Kilowatt hour, a measure of electricity delivered or consumed

L

Likelihood A measure of probability and/or level of confidence

Long-run The relatively far future

M

Market penetration The degree to which a new technology or practice enters a market for a type of equipment or service, usually measured as a percentage of sales

Market saturation The highest percentage of a market that can be captured by a type of equipment, practice, or process

Mitigation In climate change discussions, refers to actions that respond to concerns about climate change by reducing greenhouse gas emissions or enhancing sinks

O

Once-through cooling As distinct from the use of cooling towers, the practice in power plants of taking in water from a body of water (e.g., a river), using it to cool the power plant, and releasing the water back to the body of water after a single pass through the plant

P

Peaking load units Electricity supply units designed to respond to demands, often short-lived, that are significantly above normal base loads

Portfolio standards Guidelines or requirements that total electricity supply include one or more set minimums for particular sources, such as renewable energy

Power plants Facilities that produce electricity

Primary energy The amount of energy embodied in natural resources (e.g., coal, crude petroleum, sunlight) before transformation by humans. Also known as source energy

Projections Characterizations of the future, often quantitative either from extrapolations of historical trends or from models

Prospectus A formal summary of a proposed venture or project or a document describing the chief features of a proposed activity

Q

Quad Quadrillion Btus

Qualitative Characterized by units of measure that are not numerical

R

R&D Research and development

Renewable energy Energy based on resources that are naturally renewed over time periods equivalent to resource withdrawals

Risk management

Practices followed by companies and individuals to limit exposure to hazards and to limit the consequences of remaining exposure

S

Scenario A characterization of changes in the future, often associated with quantitative projections of variables of interest

Seasonal Pertaining to a season of the year, as in winter or summer

Sectors Subdivisions of a larger population, most often subdivisions of an economy such as residential, commercial, and industrial

Shell See "building shell"

Short-run The relatively near future

Simulation models Mathematical models designed to approximate the performance of a system (e.g., the energy market or the world's climate) and commonly used to quantitatively forecast elements of that system's performance

Site energy consumption The amount of energy consumed at the point of end use, not accounting for conversion losses

Solar radiation The Sun's radiant energy (in the context of this study) as deposited on the Earth in all wavelengths

Space conditioning Human interventions to modify the temperature of built spaces, including cooling and heating

Space cooling Space conditioning processes used to reduce the temperature in built spaces

Space heating Space conditioning processes used to increase the temperature in built spaces

Spatial scale Geographical size

Stakeholders Individuals, groups, and/or institutions with a stake in the outcome of a decision-making process

Statistical analysis Analyzing collected data for the purposes of summarizing information to make it more usable and/or making generalizations about a population based on a sample drawn from that population

Stochastic Characterized by risk, randomness, or uncertainty. Random or probabilistic but with some direction

Strategic Petroleum Reserve A U.S. national program and set of facilities to store petroleum as a protection against risks of supply disruptions

T

Take back A consumer reaction wherein beneficiaries of cost reductions from improvement to a technology or process undermine the improvement by using more of the improved technology or process; e.g., be setting the thermostat higher when a building is better insulated and therefore less expensive to heat

Thermal power plant A facility that produces electricity from heat

Thermoelectric See thermal power plant

Time series A series of measurements occurring over a period of time

Transient weather events Very short-lived weather happenings (e.g., thunderstorms, tornadoes) as opposed to general, long-term changes in temperature, precipitation, etc.

U

Uncertainties Unknowns that limit the completeness of an explanation or the precision and accuracy of a prediction

Urban form The physical configuration and pattern of an urbanized area

Urban heat islands The semipermanent warming of up to several degrees in urban areas compared to nearby rural areas, due to density of population, high use of energy, and prevalence of solar energy absorbing and reradiating surfaces such as concrete buildings and streets

V

Vulnerability The degree to which a system is susceptible to, or unable to cope with, adverse effects of climate change, including climate variability and extremes. Vulnerability is a function of the character, magnitude, and rate of climate variation to which a system is exposed, its sensitivity, and its adaptive capacity

In: Effects of Climate Change on Energy Production... ISBN: 978-1-60741-426-1
Editor: Alrik M. Solberg

Chapter 11

ANNEX D – ACRONYMS

U.S. Climate Change Science Program

API	American Petroleum Institute
ASHRAE	American Society of Heating, Refrigerating and Air-Conditioning Engineers
AWEA	American Wind Energy Association
BEAMS	a building energy simulation model
CALPAS	a building energy simulation model
CBECS	Commercial Building Energy Consumption Survey
CCSP	Climate Change Science Program
CCTP	Climate Change Technology Program
CDD	cooling degree days
CDM	Clean Development Mechanism
CEPA	California Environmental Protection Agency
CO_2	carbon dioxide
CRIEPI	Japanese Central Research Institute of Electric Power Industry
DD-NEMS	an energy system model based on NEMS (see below)
DOC	U.S. Department of Commerce
DOE	U.S. Department of Energy
DOE-2	building energy simulation software
DOI	U.S. Department of the Interior
DOT	U.S. Department of Transportation

EIA	Energy Information Administration
EPA	Environmental Protection Agency
EPRI	Electric Power Research Institute
FEDS	a building energy simulation model
GCM	General Circulation Model of the earth's atmosphere
GHG	greenhouse gas(es)
GOM	Gulf of Mexico
GW	gigawatt
GWh	gigawatt/hour
HADCM3	Hadley Center Coupled GCM Model, Version 3
HDD	heating degree days
HVAC	heating, ventilating, and air conditioning
ICF	an international consulting firm in Washington, DC
IEA	International Energy Agency
IGCC	integrated gasification combined cycle
IPCC	Intergovernmental Panel on Climate Change
ITC	investment tax credit
JIP	Joint Industry Program
LNG	liquefied natural gas
MMS	U.S. Minerals Management Service
MW	megawatt
NACC	U.S. National Assessment of Implications of Climate Variability and Change
NARUC	National Association of Regulatory Utility Commissioners
NCAR	National Center for Atmospheric Research
NEMS	National Energy Modeling System (Energy Information Administration)
NGCC	natural gas combined cycle
NOAA	National Oceanic and Atmospheric Administration
NRC	Nuclear Regulatory Commission
NREL	National Renewable Energy Laboratory
NSTC	National Science and Technology Council
ORNL	Oak Ridge National Laboratory
PAD	Petroleum Administration for Defense
PCM	parallel climate model
PTC	production tax credit
PV	photovoltaic
RECS	Residential Energy Consumption Survey

RGGI	Regional greenhouse gas initiative
SAP	Synthesis and Assessment Product
SAP 4.5	Synthesis and Assessment Product 4.5 (this document)
TRU	Trailer Refrigeration Units
USDA	U.S. Department of Agriculture
USGS	U.S. Geological Survey
WWF	World Wildlife Fund

INDEX

A

B

C

D

E

F

G

H

I

L

M

N

O

P

Q

R

S

T

U

V

W

Y